3rd International Conference on Structural Nano Composites (NANOSTRUC 2016)

IOP Conference Series: Materials Science and Engineering
Volume 195

Aberdeen, United Kingdom
12 – 15 September 2016

ISBN: 978-1-5108-4039-3
ISSN: 1757-8981

Printed from e-media with permission by:

Curran Associates, Inc.
57 Morehouse Lane
Red Hook, NY 12571

Some format issues inherent in the e-media version may also appear in this print version.

Printed by Curran Associates, Inc. (2017)

For permission requests, please contact the Institute of Physics
at the address below.

Institute of Physics
Dirac House, Temple Back
Bristol BS1 6BE UK

Phone: 44 1 17 929 7481
Fax: 44 1 17 920 0979

techtracking@iop.org

Additional copies of this publication are available from:

Curran Associates, Inc.
57 Morehouse Lane
Red Hook, NY 12571 USA
Phone: 845-758-0400
Fax: 845-758-2633
Email: curran@proceedings.com
Web: www.proceedings.com

Table of contents

Volume 195

3rd International Conference on Structural Nano Composites (NANOSTRUC2016)

12–15 September 2016, Aberdeen, Scotland, UK

Accepted papers received: 27 April 2017
Published online: 11 May 2017

Preface

011001
OPEN ACCESS
3rd International Conference on Structural Nano Composites (NANOSTRUC2016)

J Njuguna

011002
OPEN ACCESS
Peer review statement

Papers

012001
OPEN ACCESS
Synthesis and characterization of polyurethane/bentonite nanoclay based nanocomposites using different diisocyanates: relation between mechanical and thermal properties

Javier Bocchio, Víctor Wittemberg and Javier Quagliano.....1

012002
OPEN ACCESS
The estimation of quantitative parameters of oligonucleotides immobilization on mica surface

T I Sharipov and R Z Bakhtizin.....12

012003
OPEN ACCESS
Feasibility of Plasma Treated Clay in Clay/Polymer Nanocomposites Powders for use Laser Sintering (LS)

Alaa Almansoori, Ryan Seabright, C Majewski and C Rodenburg.....17

012004
OPEN ACCESS
Chitosan-Based Nanocomposite Beads for Drinking Water Production

ML Masheane, LN Nthunya, SS Sambaza, SP Malinga, EN Nxumalo, BB Mamba and SD Mhlanga.....23

012005
OPEN ACCESS
Asphaltenes as new objects for nanoelectronics

M Yu Dolomatov, A M Petrov, R Z Bakhtizin, M M Dolomatova, I R Khairudinov, S A Shutkova, E A Kovaleva and N Kh Paymurzina.....29

012006
OPEN ACCESS
Enhanced solar light photodegradation of brilliant black bis-azo dye in aqueous solution by F, Sm^{3+} codoped TiO_2

Sabastian S Mukonza, Edward N Nxumalo, Bhekie B Mamba and Ajay K Mishra.....33

012007
OPEN ACCESS
Synthesis, Structural and Morphological Property of $BaSnO_3$ Nanopowder Prepared by Solid State Ceramic Method

Jibi John, V.P Mahadevan Pillai, Anitta Rose Thomas, Reji Philip, Jaison Joseph, S. Muthunatesan, V. Ragavendran and Radhakrishna Prabhu.....53

012008
OPEN ACCESS
Oil Based Drilling Fluid Waste: An Overview on Environmentally Persistent Pollutants

Shohel Siddique, Lorraine Kwoffie, Kofi Addae-Afoakwa, Kyari Yates and James Njuguna.....58

012009
OPEN ACCESS
Tensile and Flexural Properties of Hybrid Graphene Oxide / Epoxy Carbon Fibre Reinforced Composites

G Watson, K Starost, P Bari, N Faisal, S Mishra and J Njuguna.....67

012010
OPEN ACCESS
Designing of Hybrid Structured Glass Laminated Transparent Nano Composites through Vacuum infusion Technique

A. Mukherji, N.D. Tarapure and G.N. Wakure.....73

012011
OPEN ACCESS
Environmental Particle Emissions due to Automated Drilling of Polypropylene Composites and Nanocomposites Reinforced with Talc, Montmorillonite and Wollastonite

K Starost, E Frijns, J V Laer, N Faisal, A Egizabal, C Elizextea, I Nelissen, M Blazquez and J Njuguna.....85

012012
OPEN ACCESS
Computational study of nanostructured composite materials for photonic crystal fibre sensors

Jincy Johny, Radhakrishna Prabhu and Wai Keung Fung.....92

012013
OPEN ACCESS
Investigations on mechanism of self-healing and cavity filling in case of steel inoculated with seashell powder

Parvaiz Habibullah, Ghazala Siddiqui and Yasir Saleem.....98

IOP Conf. Series: Materials Science and Engineering **195** (2017) 011001 doi:10.1088/1757-899X/195/1/011001

The 3rd International Conference on Structural Nano Composites (NANOSTRUC2016)

J Njuguna[1*]

[1] Centre for Advanced Engineering Materials, School of Engineering, Robert Gordon University, Aberdeen, AB10 7GJ, United Kingdom

E-mail: j.njuguna@rgu.ac.uk

Abstract. The Robert Gordon University, Aberdeen hosted The NANOSTRUC 2016 in Aberdeen (Scotland, United Kingdom). The conference focused on 'Nanoscience and Nanotechnologies – Recent Advances towards Nanoproducts and Applications'. It promoted activities in various areas on materials and structures by providing a forum for exchange of ideas, presentation of technical achievements and discussion of future directions. The conference benefitted from keynote lectures focused on topical issues in nanosciences and nanotechnology. The key sessions were on Application of Nanomaterials and Nanocomposites, Functional Nanocomposites, Graphene and Carbon-based Nanocomposites, Metallic and Metals Oxide Nanocomposites, Sustainability - Nanosafety & Environment, Toughness of Polymer Nanocomposites, Biocomposites and Nanofibres and on Fibre Reinforced Composites. A sample of papers presented at the NANOSTRUC 2016 are briefly summarised in this Issue

1. Introduction

Large quantity of nanomaterials such as carbon nanotubes, nanofibers, polyhedral oligomeric silsesquioxane (POSS), graphene, SiO_2 and layered silicates such as montmorillonite is presently available. This is made possible due to the establishment of well-developed manufacturing technologies such as chemical vapor deposition method, ball milling and electrospinning. The global nanotechnology market is expected to reach $90.5 billion by 2021 from $39.2 billion in 2016 at a compound annual growth rate (CAGR) of 18.2%, from 2016 to 2021 [1]. Growing demand for nanocomposites in industrial applications such as in transport sector, increased application of nanocomposites in building and construction, high volume consumption in electronics and semiconductors are some of the major factors that have been driving the market.

A sample of papers presented at the NANOSTRUC 2016 are briefly summarised in this Issue. The 3rd International Conference on Structural Nano Composites (NANOSTRUC2016) followed on previous conferences held in Madrid in Spain [2,3] and in Bedfordshire in the UK [4,5]. To start with, the conference benefited from Dr Clare Hoskins Keynote Lecture on 'Temperature Controlled Theranostics for Cancer Therapy' [6] highlighting the importance of nanomaterials and nanotechnology in the frontline against cancer. These cytotoxicity studies combined with cellular uptake studies showed the formulations to be significantly more effective compared with gemcitabine thanks to utilisation of hybrid nanoparticles composed of an iron oxide core surrounded by a rigid gold shell.

2. Sensors and Nanoelectronics

Dr Radhakrishna Prabhu Keynote talk covered 'Structural and nonlinear optical property of $BaSnO_3$ nanopowder prepared by solid-state ceramic method' concentrating on $BaSnO_3$ [7]. This material has a cubic perovskite-type oxide that behaves as an n-type semiconductor with a wide band gap of 3.4 eV and remains stable at temperatures up to 1000°C. It follows on that barium stannate is used in optical applications, in capacitors and ceramic boundary layers, and as a promising material to produce gas phase sensors for the detection of carbon monoxide and carbon dioxide. On the same topic on optical properties, Mukonza et al [8] investigated the effects of increasing reusability iterations on morphological, optical and crystallinity properties of trivalent-doped TiO_2/CNT photocatalysts as an alternative water treatment method. They also investigated the improvement of photocatalytic efficiency of TiO_2 during the photo-mineralisation of brilliant black (BN) bis-azo dye pollutant in aqueous solution [8]. The researchers improved the visible light activity of TiO_2 photocatalyst semiconductor through co-doping of fluorine (F) and trivalent samarium ions (Sm^{3+}) into a TiO_2 matrix using a modified sol-gel synthesis method. Further, Johny et al [9] conducted computational studies on drawback of commercially available photonic crystal fibres (PCF) which is their limited operating wavelengths, which is mostly in the infrared (IR) spectral band. Theoretical study revealed that a nanostructured PCF experiences reduced confinement losses and improved mode field diameter. On a different note, Yu et al [10] reported on the development asphaltenes as new objects for nanoelectronics. Asphaltenes are complex materials that exist in crude oil, bitumen and high-boiling hydrocarbons distillates. They noted that the paramagnetic phase of asphaltenes is organic amorphous wide band gap semiconductor.

3. Water-Nanomaterials Research

Water research extensively utilise nanomaterials for the treatment, remediation, and pollution prevention. The focus of the lecture will to provide an overview of the nanoscale materials for wastewater chemistry as futuristic materials. Prof. Ajay Mishra Keynote Lecture covered this topic on a talk titled 'Nanoscale Sovereignty in Waste Water Chemistry' [11]. It specifically focused on nanomaterials in water research for the treatment, remediation, and pollution prevention. Masheane et al [12] studied chitosan-alumina/functionalised multiwalled carbon nanotube (f-MWCNT) nanocomposite beads to developed and investigate the reduction of various physico-chemical parameters for drinking water. The researchers reported that the chitosan nanocomposite beads offer a promising alternative material for reduction and removal of various physico-chemical parameters for production portable water. In the same field, Prof Sabelo Mhlanga's Keynote presentation focused on 'Chitosan-Based Nanocomposite Beads for Drinking Water Production' [12]. His work covered the chitosan-alumina/functionalised multiwalled carbon nanotube (f-MWCNT) nanocomposite beads that were developed and investigated for the reduction of various physicochemical parameters from water samples collected from open wells used for drinking purposes by a rural community in South Africa.

4. Hybrid Composites

Lab-scale preparation of polymer nanocomposites with layered silicates developed in the last two decades and the most important factors that contribute to the structure of polymer-based clay nanocomposites identified. However, most of the experiments are laboratory-scale and the question how to up-scale the process toward industrial production scale still remains not fully answered. Another issue that requires more research attention is recycling of polymer nanocomposites. Among different methods, pyrolysis seems to be an interesting alternative. This study was the focus of a Keynote Lecture titled 'Polymer Nanocomposites with Layered Silicates - Up- Scaling of Production and Recycling' by Prof. Pielichowski [13]. His team of researchers have also developed a novel method of polypropylene/montmorillonite nanocomposites recycling by pyrolysis to produce pyrolysed montmorillonite (MMT) for the preparation of polymer composites.

An interesting paper by Almansoori et al [14] reported on the fabrication and characterization of plasma-treated organoclay/Nylon 12 nanocomposite carried out with the aim of achieving better dispersion of clay platelets on the Nylon 12 particle surface. They used the air-plasma etching to enhance the compatibility between clays and polymers to ensure a uniform clay dispersion in composite powders. Meanwhile, Starost et al [15] studied the effect on nanoparticle emissions due to drilling on polypropylene (PP) reinforced with 20% talc, 5% montmorillonite and 5% wollastonite (WO). The study is the first to explore the nanoparticle release from WO and talc reinforced composites and compares the results to previously researched MMT in the literature. The size distribution data displayed a substantial percentage of the particles released from the PP, PP/WO and PP/MMT samples to be between 5-20 nm, whereas the PP/talc sample emitted larger particle diameters. The study noted that the high usage of nanomaterials and nanotechnology along with the environmental impact of certain materials used in nanocomposites manufacturing need due considerations.

Hierarchically structured hybrid composites are ideal engineered materials to carry loads and stresses due to their high in-plane specific mechanical properties. Watson et al [16] reported on the tensile and flexural properties of hybrid graphene oxide /epoxy carbon fibre reinforced composites (GO/EP/CFRC). It was observed that the tensile and flexural strength of the EP/CF decreased with the addition of GO but increased with GO weight concentration in the nanocomposites studied. Mukherje et al. [17] reported on a new invention that comprises a method of making a transparent glass laminated nanocomposite product. The product contains a bidirectionally oriented E-Glass fabric an essentially bidirectional yarn woven fabrics is stretched bidirectionally by specially fabricated steel frame associated with both co- and counter-rotating device.

5. Green Composites

Dr Youssef Habibi delivered another interesting Keynote lecture titled 'Nanocelluloses: Emerging Biobased Building Blocks for Nanostructured Materials' [18]. It also provided few examples related to nanocellulose processing using various techniques to access useful nanostructured materials for different applications. This interesting lecture. Sustainable Bioplastics from Microbes and Waste by Dr Christine Edward reported on strategies for exploiting microbes grown on waste substrates for production of polyhydroxalkanoates (PHAs), a group of polyesters produced mainly as storage compounds in bacteria where biosynthesis triggered by nutrient limitation and stress [19].

A study by Habibullah et al [20] investigated on the mechanism of self-healing and cavity filling in case of steel inoculated with seashell powder. It follows that the seashells such as the abalones use a protein mortar for reinforcement. The researchers identified the protein and mechanism how the protein mortar stretches itself into ligaments that bridge the gap, with the help of microscopy. Sharipov and Bakhtizin [21] work developed a method for estimation of quantitative parameter of immobilization of oligonucleotides is reliant on on their degree of aggregation depending on the fixation conditions on the surface of mica. The study reports on aggregation of oligonucleotides immobilized on mica surface morphology and associated physic-chemical features. Another keynote lecture by Prof Kalia Susheel titled 'Biocomposites and Their Applications' proposed that the biografting of natural organic molecules on natural fibres is one of the best environmentally friendly methods [22]. Applications of such green composite materials include biomedical applications, structural applications, consumer products, food packaging, transportation, and textile, sports & leisure industry.

6. Oil and Gas Applications

The Keynote Talk on 'Nanotechnology Needs in Oil and Gas Industry Applications' by Prof Babs Oyeneyin covered the potential for nanotechnology application across the supply chain of the oil industry operations from exploration through to well construction and reservoir management/monitoring to production flow assurance [23]. He delivered show case examples of key areas of potential applications. The talk noted that the industry key challenge is to reduce cost and be more operationally efficient to guarantee the oil/gas flow assurance and that the nanotechnology is one of the key technologies required to achieve these objectives. On the same topic, Siddique et al [24] review study provided an overview on the basic concepts of drilling fluids and their functions, sources and characterisation of drilling wastes, and highlights some environmentally significant elements including different minerals present in drilling waste stream. The study is part of an on-going work at Robert Gordon University, Aberdeen, to convert the drilling fluid waste into polymer additives for oil and gas nanocomposites applications.

7. Best Student Poster Awards

Wiley and Nanoscale Horizons, Royal Society of Chemistry sponsored the young researchers' awards for NANOSTRUC 2016. The poster titled 'The Effect of Talc, Montmorillonite (MMT) and Wollastonite (WO) on Nanoparticle Release due to Mechanical Drilling from Polypropylene (PP) Composites' [15] by Kristof Starost won the Best Poster Award supported by the Nanoscale Horizons, Royal Society of Chemistry. The study provided new insight and a better understanding on how various fillers effect the release of nanoparticles during machining of composites to evaluate and quantify the full exposure risks associated to nanorelease and nanoparticle at this stage of manufacturing process. The second best poster award went to Ekaterina Mitrofanova poster on her work on 'Dimensional Dependent Electronic Properties in Tetragonal Iron Chalcogenides' [25]. The study investigated a change of electronic structure during the transition from bulk material to two-dimensional one for iron chalcogenides (FeSe, FeTe, for up to 15 layers and bulk material) with ab initio calculations. Nanoscale Horizons, Royal Society of Chemistry, also sponsored the Second Poster Award.

Karsta Heinze and Chinere Okolo both received the Third Poster Prize Award sponsored by Wiley. Heinze work focused on 'The Wheat Grain as a Cemented Granular Material: Nanoscale Investigation of its Mechanical Properties' [26]. She applied a multiscale experimental and modelling approach to investigate the nanomechanical properties of the composing polymers and the way they influence grain fragmentation, with the aim of linking nanomechanics of specific grain samples to their milling behaviour, which was determined with a laboratory mill nanoscale and macroscale mechanical properties. On the other hand, Okolo studied 'Dynamic Mechanical Analyses for Molecular Level Engineering of Advanced Subsea Polymers' [27]. The poster presented a study on dynamic mechanical analysis (DMA) characterization of engineering polymer with focus on its mechanical behaviour and thermal stability at elevated temperatures. She noted that challenging subsea field developments in ultra-deepwaters has significantly pushed the need for reliable engineering materials and products for high pressure and high temperature applications, hence the need for in-depth knowhow on nanoscale mechanics.

Acknowledgments

We would like to acknowledge the hard work, professional skills and efficiency of the team, which ensured the general organisation, in particular the NANOSTRUC 2016 Scientific and Organising Committees, Session Chairs and all assistants. We would also like to acknowledge support from our industrial and Media partners – Wiley, BREC and Nanoscale Horizons, Royal Society of Chemistry; UK's Knowledge Transfer Network- Nanotechnology (UK); Innovation (Greece); The Minerals, Metals & Materials Society (TMS); Nafigate; VisitAberdeen; LIFE 11 programmes through SIRENA project. Financial funding from UK-India Research Initiative (Project No. DST/INK/UK/P-108/2014) on graphene hybrid composites is acknowledged. Special thanks to our host Robert Gordon University and academic partners including Consejo Superior de Investigaciones Científicas (CSIC) and University of South Africa.

References

[1] Market Research Reports, NAN031G, 2016.
[2] Njuguna, J., and R. Verdejo, 2014, *IOP Conf. Ser.: Mater. Sci. Eng.* **64**, No. 1. IOP Publishing.
[3] Njuguna, J, Verdejo R and Starost K, 2014, NANOSTRUC 2014Book of Abstracts.

[4] Njuguna J and Sachse S, 2012 NANOSTRUC 2012 Book of Abstracts.
[5] Njuguna J 2012 *IOP Conf. Ser.: Mater. Sci. Eng.* **40** 01100, IOP Publishing, 2012.
[6] Hoskins C, 2016, NANOSTRUC 2016 Book of Abstracts, pg 16.
[7] John J, Pillai VPM, Thomas AR, Philip R and Prabhu R, 2016, NANOSTRUC Book of Abstracts, pg 27.
[8] Mukonza SS, Nxumalo EN, Mamba BB and Mishra AK, 2017, *IOP Conf. Ser.: Mater. Sci. Eng*. **XX**, No. XX. IOP Publishing.
[9] Johny J, Fung WK and Prabhu R, 2017, *IOP Conf. Ser.: Mater. Sci. Eng*. **XX**, No. XX. IOP Publishing.
[10] Yu DM, Bakhtizin RZ, Petrov AM, Shutkova SV, Dolomatova MM, Hayrudinov IR, 2017, *IOP Conf. Ser.: Mater. Sci. Eng*. **XX**, No. XX. IOP Publishing.
[11] Mishra AK, 2016, NANOSTRUC Book of Abstracts, pg 17.
[12] Masheane ML, Nthunya LN, Sambaza SS, Malinga SP, Nxumalo EN, Mamba BB and Mhlanga SD, 2017, *IOP Conf. Ser.: Mater. Sci. Eng*. **XX**, No. XX. IOP Publishing.
[13] Pielichowski K, Majka T, 2016, NANOSTRUC Book of Abstracts, pg 30.
[14] Almansoori A, Seabright R, Majewski C and Rodenburg C, 2016, NANOSTRUC Book of Abstracts, pg 51.
[15] Starost K, Frijns E, van Laer J, Faisal N, Egizabal A, Elizextea A, Nelissen I, Blazquez M and Njuguna, , 2017, *IOP Conf. Ser.: Mater. Sci. Eng*. **XX**, No. XX. IOP Publishing.
[16] Watson G, Starost K, Bari P, Faisal N, Mishra S and Njuguna J, 2017, *IOP Conf. Ser.: Mater. Sci. Eng*. **XX**, No. XX. IOP Publishing.
[17] Mukherji A, Tarapure ND, Wakure GN, 2017, *IOP Conf. Ser.: Mater. Sci. Eng*. **XX**, No. XX. IOP Publishing.
[18] Habibi Y, 2016, NANOSTRUC Book of Abstracts, pg 31.
[19] Edwards C and Lawton L, 2016, NANOSTRUC Book of Abstracts, pg 44.
[20] Habibullah P, Siddiqui S, Siddiqi SA, 2017, *IOP Conf. Ser.: Mater. Sci. Eng*. **XX**, No. XX. IOP Publishing.
[21] Sharipov TI and Bakhtizin RZ, 2017, *IOP Conf. Ser.: Mater. Sci. Eng*. **XX**, No. XX. IOP Publishing.
[22] Susheel K, 2016, NANOSTRUC Book of Abstracts, pg 56.
[23] Oyeneyin B, 2016, NANOSTRUC Book of Abstracts, pg 53.
[24] Siddique S, Yates K and Njuguna J, *IOP Conf. Ser.: Mater. Sci. Eng*. **XX**, No. XX. IOP Publishing.
[25] Mitrofanova Eand Mitrofanov A, 2016, NANOSTRUC Book of Abstracts, pg 85.
[26] Heinze K, Delenne JY, George M and Lullien-Pellerin V, 2016, NANOSTRUC Book of Abstracts, pg 57.
[27] Okolo C, Wood C and Inam F, 2016, NANOSTRUC Book of Abstracts, pg 91.

IOP Conf. Series: Materials Science and Engineering **195** (2017) 011002 doi:10.1088/1757-899X/195/1/011002

Peer review statement

All papers published in this volume of *IOP Conference Series: Materials Science and Engineering* have been peer reviewed through processes administered by the proceedings Editors. Reviews were conducted by expert referees to the professional and scientific standards expected of a proceedings journal published by IOP Publishing.

IOP Conf. Series: Materials Science and Engineering **195** (2017) 012001 doi:10.1088/1757-899X/195/1/012001

Synthesis and characterization of polyurethane/bentonite nanoclay based nanocomposites using different diisocyanates: relation between mechanical and thermal properties

Javier Bocchio[1], Víctor Wittemberg[2] and Javier Quagliano[1,3]

[1] División Síntesis Orgánica - Laboratorio de Elastómeros, Departamento Química Aplicada, Instituto de Investigaciones Científicas y Técnicas para la Defensa (CITEDEF), Av. Juan Bautista de Lasalle 4397, Buenos Aires, Argentina.
[2] División Química Analítica, Departamento Química Aplicada, Instituto de Investigaciones Científicas y Técnicas para la Defensa (CITEDEF), Av. Juan Bautista de Lasalle 4397, Buenos Aires, Argentina.
[3] E-mail: jquagliano@citedef.gob.ar

Abstract. Polyurethanes (PUs) and polyurethane nanocomposites (PUNC) with bentonite nanoclay were prepared by the reaction of toluene-2,4-diisocyanate (TDI), dimeryl diisocyanate (DDI) and isophorone diisocyanate (IPDI) with two different polymers: hydroxyl terminated polybutadiene (HTPB) and polytetramethylene ether glycol (PTMEG), and the chains were further extended with 1,4-butanediol (1,4-BDO) to get final PUs and PUNCs. PUNCs were prepared by dispersing within the polymers a commercial and a synthesized bentonite nanoclay by mechanical dispersion. Mechanical properties showed that the addition of a small amount of nanoclay resulted in a significant increase in tensile strength and reduction in elongation at break (maximum increase of 2.3 and 5-times reduction, respectively, for a HTPB-TDI-BDO PUNCs). Thermal analysis revealed that the addition of nanoclays improved the thermal stability and increased decomposition temperature of PUNCs. We concluded that there is a positive correlation between mechanical and thermal properties as a result of nanoclay addition.

1. Introduction

PUs have found an extensive use in numerous commercial applications such as coatings, foams, adhesives, sealants, synthetic leathers, membranes, elastomers, as well as in many biomedical applications [1] [2]. Considerable attention has been paid in the literature to develop an understanding of the structure–property relationships of polyurethanes (PUs), which has acquired significance because of the broad range of applications of these materials.

Due to the diversity of combinations of flexible and rigid segments, materials can be formulated with different properties, from soft and flexible elastomers to hard thermoplastics with a high modulus [3]. Rigid or hard segments in polyurethanes consist of diisocyanate residues, chain extenders and urethane groups. The more symmetric the molecules of diisocyanate and the chain extenders, the easier will the hard segments crystallise, and the higher will be the tensile strength.

If the hard segments contain bulky aromatic or cycloaliphatic groups, the rigidity will increase, resulting in interchain steric hindrance. In the case of polybutadiene, the soft segments consist of flexible non crystalline polymer chains.

HTPB, mixed with BDO and cured with the diisocyanates HDI, DDI, CHDI, MDI and TDI has been studied [4]. Isophorone diisocyanate (IPDI) was used to cure HTPB [5].

Polymer-layered silicate nanocomposites have gained a great deal of attention due to their superior properties like high dimensional stability, high heat deflection temperature, reduced gas permeability, improved flame retardancy, and enhanced mechanical properties compared to pure polymers [6]. The nanoscale clay particles can be dispersed to a scale of few nanometers in the polymer matrix, and such dispersions are produced by shear or chemical reactions or a combination of both [7]. In a previous contribution [8] we have shown that an increase in decomposition temperature was observed when nanoclay was loaded at 3,5% onto an elastomeric PU made from TDI, PTMEG and BDO, while not when nanoclay content was lower (1,5%). Increasing of the decomposition temperature was attributed to the reduced diffusion of decomposition products within the polymer matrices [17]. In this contribution, we prepared PUNCs based in two different polyols and cured with TDI, DDI and IPDI and chain extended with 1,4-BDO, and their mechanical and thermal properties were studied to find correlations.

2. Materials and Methods

Reactives: a polyether polyol poly(tetramethylene)glycol (Voranol®, mean molecular weight 2435 g/mol, hydroxyl index 56), was obtained from Mentvil Argentina S.A, Buenos Aires. HTPB was from chinese origin, with a M_w of 4300 Da and a polydispersity index of 1,4. An isomeric mixture of tolydene diisocyanate (TDI) was utilized as received from Petroquímica Río Tercero S.A, Córdoba, Argentina. Isophorone diisocyanate (IPDI) at 98% was purchased from Aldrich Co and DDI was purchased from Henkel Corporation. The reactive nanoclay Cloisite C30B containing 90 mequiv./100 g clay of quaternary ammonium ions was donated by Southern Clay Products, US. The quaternary ammonium ion has the structure $N+(CH_2CH_2OH)_2(CH_3)T$, with T representing an alkyl group of approximately 65% $C_{18}H_{37}$, 30% $C_{16}H_{33}$, and 5% $C_{14}H_{29}$. A N-dodecil quinolinium-interchanged bentonite (Nanoclay*) was prepared by intercalation of natural bentonite (Lake Pellegrini bentonite, Patagonia, Argentina, provided by National Mining Service Laboratory, Segemar) for 2 hours at 80°C and then separating the supernatant by filtering. Differential scanning calorimetry (DSC) studies were performed in a Q20 apparatus (TA Instruments), using 5 mg samples in opened heating pans, at a heating rate of 10°C/min.

3. Synthesis of polyurethanes

The polyols, PTMEG and HTPB, were dried under vacuum for one hour. PTMEG or HTPB were allowed to react with a calculated quantity of the diisocyanate viz. TDI, DDI or IPDI at room temperature in a three-necked reaction kettle with a mechanical stirrer under vaccum (4 mm Hg). The diisocyanate was added dropwise from the dropping funnel over a period of 15 min. The catalyst dibutyl tin laureate (DBTL) was added to the polyol in the beginning itself. The mixture was stirred continuously for about 30-45 min to obtain NCO terminated prepolymer. This was allowed to react further with the required amount of the crosslinker for about 15 min or until the reaction mixture became viscous. This was followed by degassing after which the mixture was poured into preheated teflonated moulds. The curing was allowed to proceed at room temperature for 7 days. After cooling the crosslinked polyurethanes were used for testing. For preparing PUNCs, nanoclay was dispersed in the correspondent polymer by mixing with a Cowles blades at 3000 rpm for 1 minute. The composition of the polyurethanes synthesised and their number codes are given in Table 1.

Table 1. Polyurethanes and polyurethane nanocomposites (PUNCs) prepared from hydroxy-terminated polybutadiene (HTPB) and poly-tetramethylene glycol (PTMEG) cured with 2 different isocyanates and chain extender with 1,4-butanediol (BDO).

Sample	PTMEG	HTPB	TDI	DDI	IPDI	BDO	Nanoclay, % w/w
1	60	-	28,2	-	-	7,2	-
2	60	-	28,2	-	-	7,2	N-dodecil quinolinium-interchanged bentonite, 1
3	60	-	28,2	-	-	6,9	Cloisite 20 A, 1
4	15	-	-	-	5	1,5	-
5	15	-	-	-	5	1,5	Cloisite 20 A, 1
6	-	69,4	6,24	-	-	3,72	-
7	-	38,2	-	-	3,6	0,7	-
8	-	69,4	6,24	-	-	3,72	Cloisite 20 A, 1
9	-	38,2	-		3,6	0,7	Cloisite 20 A, 1
10	-	179.38	-	68.70	-	10.39	-
11	-	179.50	-	66.24	-	10.44	Cloisite 20 A, 1

4. DSC Studies

In general, the process of PU decomposition normally begins at a temperature between 200 and 370°C. The emerging isocyanates from decomposition react among themselves to form dimers and trimers, which in turn, in a second stage, decompose to produce substituted ureas and isocyanurates. This second stage occurs in a range of temperatures between 350 and 500°C. At higher temperatures than 500°C, products formed completely decompose to give volatile material and a carbonaceous residue (third stage) [9].

The thermal degradation temperature of isocyanurate rings is ca. 300–340°C, while in the case of traditional polyurethanes, it does not exceed 250°C [2]. This may suggest that our PU have a certain degree of isocyanurate chemicals bonds in its structure.

4.1 Polyurethanes with PTMEG

There have been proposed three mechanisms for the decomposition of urethane bonds: dissociation to isocyanate and alcohol, formation of primary amine and olefin, and formation of secondary amine and carbon dioxide. These reactions may proceed simultaneously or separately [10].

Generally, polyurethanes decompose in three steps and the temperature which the weight loss is maximum in the range of 380-500 °C. In glycerine containing PUs, ramified structures are generated, increasing the amount of flexible chains and lead to relative stability, because the possibility of ordering the main molecular chain is increased. The first stage was associated with the hard segment degradation and the second one, with the soft segment [11].

In Table 2 we summarize the main findings about the thermal behaviour of the PU and PUNCs synthesized:

Table 2. Thermal behaviour of the PU and PUNCs synthesized with PTMEG.

T_1: Temperature of the first exothermic peak
T_2: Temperature of the second exothermic peak

Sample	T_1	T_2
1-PTMEG-TDI-BDO	387.07	408.34
2-PTMEG1% Nanoclay* -TDI-BDO	382.20	415.11
3-PTMEG1% C20A –TDI-BDO	390.96	431.36
4-PTMEG-IPDI-BDO	365.19	419.94
5-PTMEG 1% C20A-IPDI-BDO	365.70	409.76

In the samples with PTMEG-TDI-BDO we can observed that the incorporation of two types of nanoclays considerably increased the decomposition temperatures of the second stage (7 ºC and 23 ºC) and this results are coherent with the work of Yun, Tien and Kung Hwa Wei [12], who studied the influence of the type of nanoclays (with 1OH, 2OH and 3OH) and observed that PU with nanoclay with higher content of OH groups improved thermal stability. Results are shown in Figure 1.

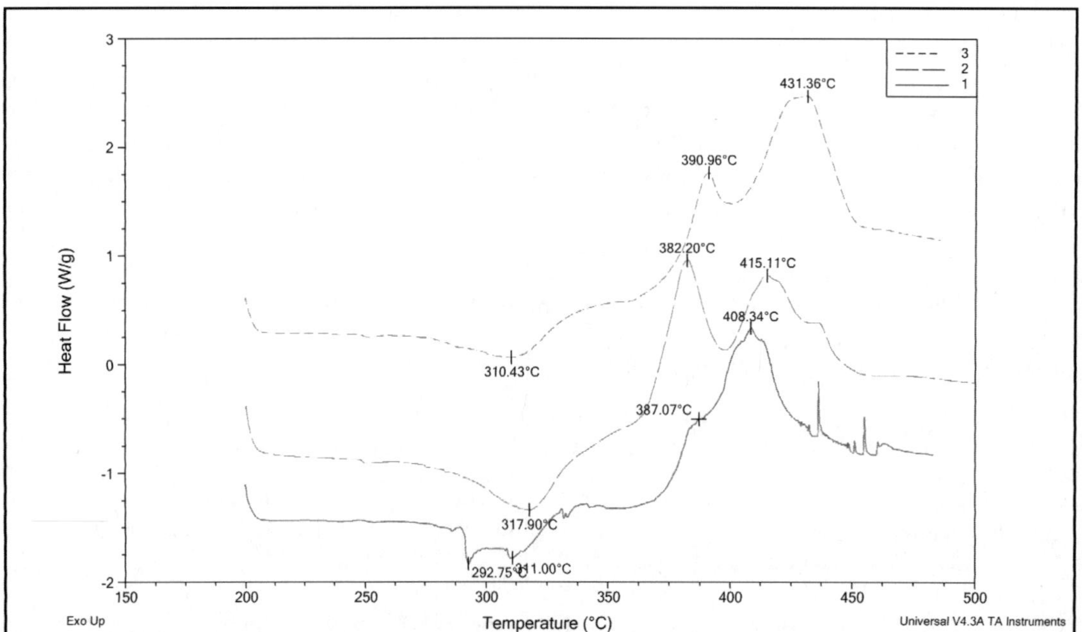

Figure 1. Differential scanning calorimetry (DSC) thermograms for 1: PTMEG-TDI-BDO, 2:PTMEG1%+Nanoclay*-TDI-BDO and 3: PTMEG+1% C20A –TDI-BDO.

In the samples with PTMEG-IPDI-BDO we observed that the incorporation of nanoclay did not affect the first exotermic peak, and also did not improve thermal stability in the second peak. On the other side, Saha et al. [13] observed that the addition of a similar nanoclay in a PU based on PTMEG-IPDI-BDO improved the thermal stability, although in that case with higher percentages than 1%.

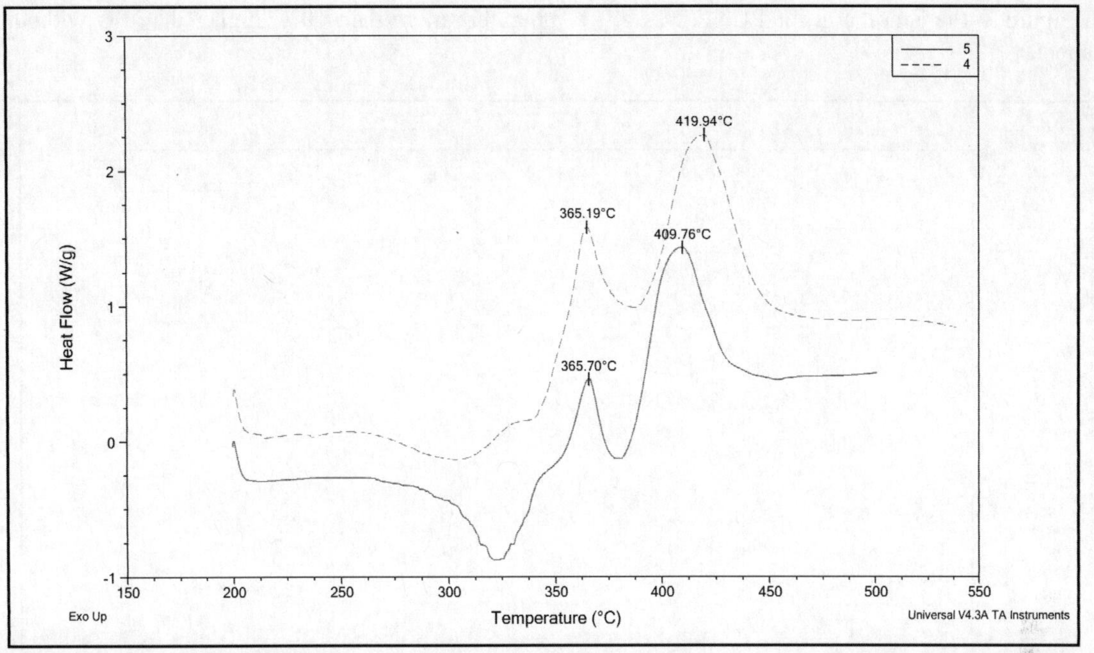

Figure 2. Differential scanning calorimetry (DSC) thermograms for 4: PTMEG-IPDI-BDO and 5: PTMEG+1% C20A-IPDI-BDO.

4.2 Polyurethanes with HTPB

The decomposition of PU with HTPB occurs in two stages, one exothermic and the other endothermic. The initial exotherm results from the positive energy balance of the endothermic depolymerization process and the exothermic cyclization and cross-linking process of the HTPB that has not undergone depolymerization. The second process is mostly the result of depolymerization of the cross-linked residue formed in the cyclization and cross-linking reactions of the first decomposition step [14].

In Table 3 we summarize the main findings about the thermal behaviour of the PU and PUNCs synthesized:

Table 3. Thermal behaviour of the PU and PUNCs synthesized with PTMEG.

T_1: Temperature of the exothermic peak
T_2: Temperature of the endothermic peak

Sample	T_1	T_2
6-HTPB-TDI-BDO	370.85	454.83
7-HTPB-IPDI-BDO	368.84	456.52
8-HTPB 1% C20A-TDI-BDO	372.85	465.46
9- HTPB 1% C20A-IPDI-BDO	373.40	468.04
10- HTPB-DDI-BDO	383.42	-
11- HTPB 1% C20A-DDI-BDO	374.73	-

In Figure 3, the endotherm for PUNC was at a temperature more than 10°C higher than PU without nanoclay:

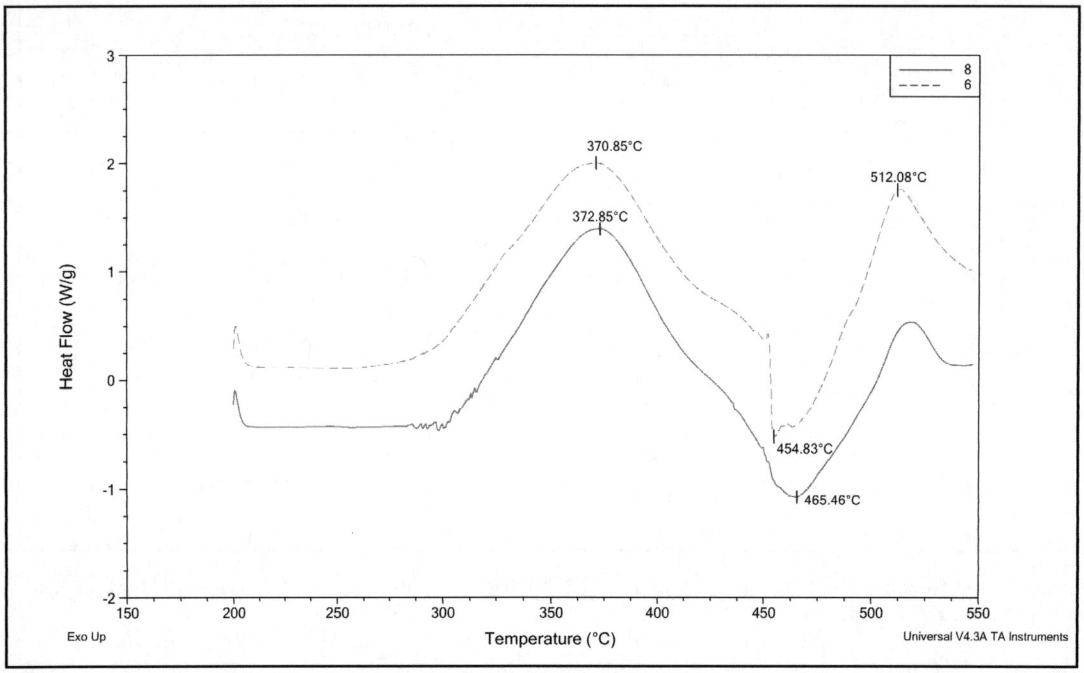

Figure 3. Differential scanning calorimetry (DSC) thermograms for 6:HTPB-TDI-BDO and 8: HTPB 1%+C20A-TDI-BDO.

In Figure 4, same behaviour was observed for the endotherm of PUNC cured with IPDI. This suggested that the effect was not dependent on the curative utilized but could be explained because of the addition of nanoclay. In the samples with TDI and IPDI we could also observe (apart for the temperature increase of the endothermic peak (10,5ºC for TDI and 11,5ºC for IPDI) that the PUNC with 1% nanoclay increased the temperature of the second exothermic peak (2ºC for TDI and 4,5ºC for IPDI).

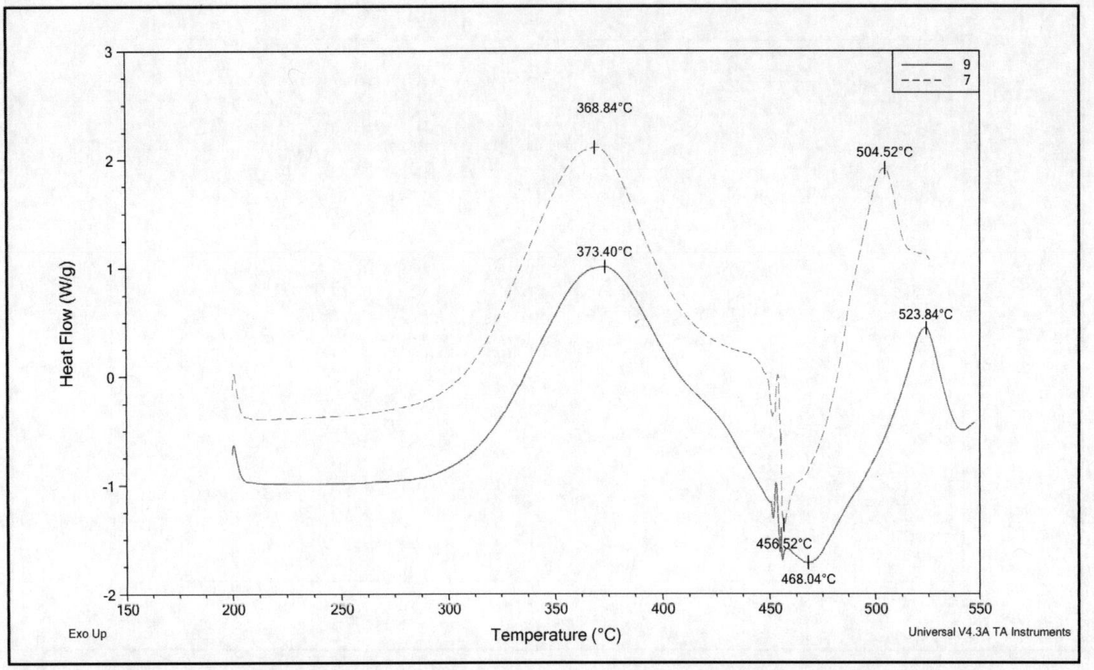

Figure 4. Differential scanning calorimetry (DSC) thermograms for 7: HTPB-IPDI-BDO and 9: HTPB+1% C20A-IPDI-BDO.

In Figure 5, PUs cured with DDI did not show the same effect as with TDI and IPDI, as a three broad exotherms were seen at around 370-380ºC, 430-440ºC and 490-500ºC. The addition of nanoclay decreased the temperature for the first exothermic peak (8,5ºC) and no endotherm could be observed. This effect could be explained by the fact of DDI is an aliphatic isocyanate with a high M_W: other studies showed that a larger portion of layered silicates is distributed in the hard-segment phase when aliphatic isocyanate is used as curative [12]. Then interaction between different aliphatic chains are weakened by the presence of nanoclay, because their bonds are weaker compared to aromatic isocyanates or even aliphatic ones with lateral chains of low molecular weight (such as those of IPDI).

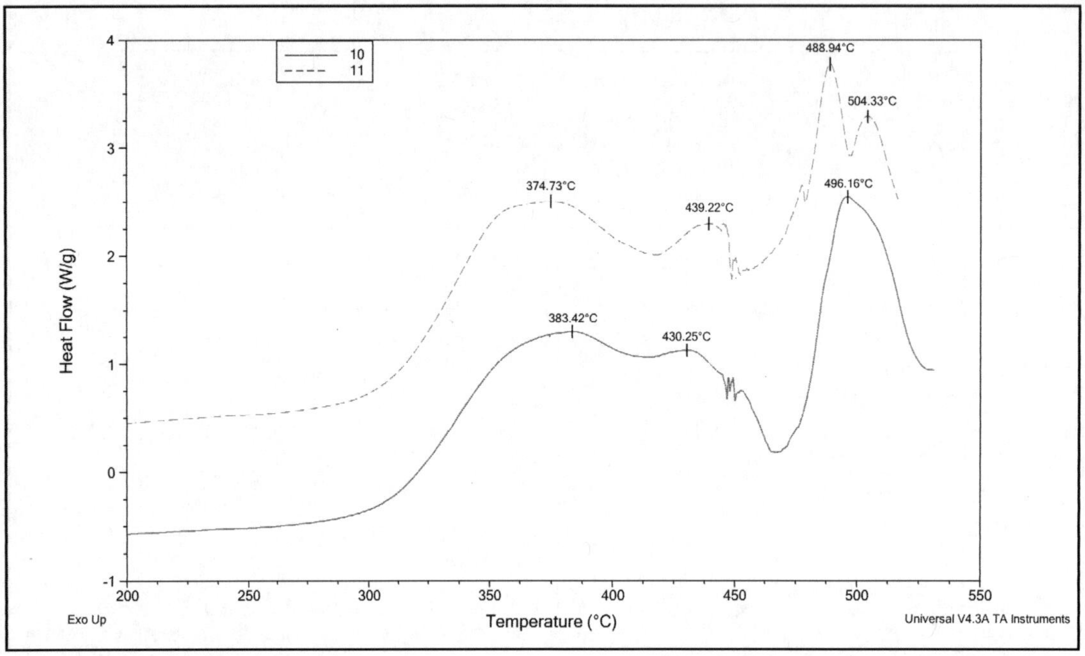

Figure 5. Differential scanning calorimetry (DSC) thermograms for 10: HTPB-DDI-BDO and 11: HTPB+1% C20A-DDI-BDO.

5. Mechanical properties

The results from stress-strain and Shore hardness essays performed on the prepared samples are shown in table 4.

Table 4. Mechanical properties of the synthtesized polyurethanes and polyurethane nanocomposites (PUNCs).

Sample	Tensile strength (MPa)	Strain at maximum load (%)	Shore A Hardness
1-PTMEG-TDI-BDO	0,45	270	90
2-PTMEG1% Nanoclay* -TDI-BDO	4	105	95
3-PTMEG1% C20A -TDI-BDO	4	105	89
4-PTMEG-IPDI-BDO	0,08	900	70
5-PTMEG 1% C20A-IPDI-BDO	0,12	240	75
6-HTPB-TDI-BDO	0,09	800	55
7-HTPB-IPDI-BDO	0,1	900	60
8-HTPB 1% C20A-TDI-BDO	0,21	160	40

IOP Conf. Series: Materials Science and Engineering **195** (2017) 012001 doi:10.1088/1757-899X/195/1/012001

9- HTPB 1% C20A-IPDI-BDO	0,14	170	25
10- HTPB-DDI-BDO	0.4	390	
11- HTPB 1% C20A-DDI-BDO	0.12	950	

Table 4: Mechanical properties of the synthesized polyurethanes and polyurethane nanocomposites (PUNCs).

It was shown that the addition of 1% nanoclay resulted in an increase of nearly 10 times on the tensile strength and a three-fold reduction in strain at maximum load for PTMEG-TDI polyurethanes extended with 1,4-BDO. There were no differences between a laboratory-prepared nanoclay (sample No 2) and a commercial Cloisite nanoclay PUNC (sample No 2). Therefore, Cloisite was used for preparing other PUNCs. When IPDI was used as curing agent for PTMEG, nanoclay had a significant effect mostly in elongation at break (4-fold increase) while tensile at break diminished only around 30%.

When HTPB was used as polyol the addition of nanoclay doubled tensile at break and reduced elongation 5 times for PUNC cured with TDI. On the other hand, the utilization of the aliphatic isocyanate IPDI rendered an increase of more than 4 times in elongation at break, while tensile strength was not significantly affected (0.1 and 0.14%, respectively). In the case of PTMEG polyol cured with DDI we observed that the effect of nanoclay on tensile strength was low, possibly because the interactions among the long aliphatic chain of DDI are obstructed by nanoclay. However, elongation was more than doubled, indicating that chains are not fully restricted in their motion by nanoclay interaction. It was pointed out that mechanical bonding on the surface of nanoparticles and the stress field around is a result of shear mixing, which results in nanoclay load bearing properties [15]. Respect to mechanical resistance, polymer matrices also influence mechanical resistance through craze initiation above a critical matrix ligament thickness [16]. This highlights the importance of good dispersion of nanoclay within polymers, in order to obtain an exfoliated nanoclay and avoid the formation of crazes, fostering a high crack initiation energy.

6. Conclusions

Our results suggested that mechanical properties are correlated with thermal properties for the elastomeric materials studied. PUNC cured with TDI and IPDI displayed an increase in tensile strength with the expectable reduction in strain at maximum load. At the same time, decomposition temperature (second exotherm peak) augmented, possibly due to the reduced diffusion of decomposition products within the polymer matrices, as nanoclays are known to reduce gas permeability. The DSC results showed a slight increase of the melting temperature with Cloisite loading. The study of the thermal stability confirmed that the nanocomposites are thermally more stable than the unmodified systems. A significant increase in mechanical properties was obtained even at low percentages of nanoclay (1%) compared to the usual amounts incorporated in PUNCs (3 to 5%).

IOP Conf. Series: Materials Science and Engineering **195** (2017) 012001 doi:10.1088/1757-899X/195/1/012001

7. References

[1] Prisacariu C 2011 *Polyurethane Elastomers: From Morphology to Mechanical Aspects.* (Wien; New York: Springer).

[2] Piszczyk L, Strankowski M, Danowska M, Hejna A, Haponiuk J 2014 Rigid polyurethane foams from a polyglycerol-based polyol. *European Polymer Journal.* **57**: p. 143-150.

[3] Pizzatto L, Lizot A, Fiorio R, Amorim C, Machado G, Giovanela M, Zattera A, Crespo J 2009 Synthesis and characterization of thermoplastic polyurethane/nanoclay composites. *Materials Science and Engineering: C.* **29**(2): p. 474-478.

[4] Wingborg N 2004 *Improving the mechanical properties of composite rocket propellants.* Tesis AKADEMISK AVHANDLING, Department of Fibre and Polymer Technology, Royal Institute of Technology; (Stockholm, Sweden).

[5] Wingborg N 2002 Increasing the tensile strength of HTPB with different isocyanates and chain extenders. *Polymer Testing.* **21**(3): p. 283-287.

[6] Gintert M, Jana S, Miller S 2007 A novel strategy for nanoclay exfoliation in thermoset polyimide nanocomposite systems. *Polymer.* **48**(14): p. 4166-4173.

[7] Park J, Jana J. 2003 Mechanism of exfoliation of nanoclay particles in epoxy-clay nanocomposites. *Macromolecules.* **36**(8): p. 2758-2768.

[8] Quagliano J, Bocchio J. 2014 *Effect of nanoclay loading on the thermal decomposition of nanoclay polyurethane elastomers obtained by bulk polymerization.* 2nd International Conference on Structural Nano Composites (Nanostruct 2014). In IOP Conf. Series: Materials Science and Engineering **64**; p. 012035.

[9] Nanclares J, Gil Pilonieta A, Jaramillo F. 2010 Synthesis of polyurethanes from higuerilla oil and study of its termal degradation. (In IBEROMET XIX CONAMET/SAM; November 2-5, Viña del Mar, Chile).

[10] Gaina C, Ursache O, Gaina V, Varganici C D. 2013 Thermally reversible cross-linked poly(etherurethane)s. *eXPRESS Polymer Letters*; **7**(7): p. 636-650.

[11] Vlad S, Ciobanu C, Macocinschi D, Filip D, Spiridon I. 2009 Evaluation of some polyetherurethane elastomers for chemicals, oils and solvents resistance. *Journal of Optoelectronics and Advanced Materials.* **11**(8): p. 1160 - 1168.

[12] Tien Y I, Wei K H. 2002 The Effect of Nano-Sized Silicate Layers from Montmorillonite on Glass Transition, Dynamic Mechanical, and Thermal Degradation Properties of Segmented Polyurethane. *Journal of Applied Polymer Science.* **86**: p. 1741-1748.

[13] Saha C, Tapan K, Chaki K, Singha N. 2013 Synthesis and characterization of elastomeric polyurethane and PU/clay nanocomposites based on an aliphatic diisocyanate. *J. Appl. Polymer Sci.* **130**(5): p. 1-7.

[14] Chen J K, Brill T B. 1991 Chemistry and Kinetics of Hydroxyl-terminated Polybutadiene (HTPB) and Diisocyanate-HTPB Polymers during Slow Decomposition and Combustion-like Conditions. *Combustion and Flame.* **87**: p. 217-232.

[15] Wu S. 1998 A generalized criterion for rubber toughening: the critical matrix ligament thickness. *Journal of Applied Polymer Science*, 35, p. 544–553.

[16] Nayani M, Gunashekar S, Abu-Zhara N. 2013 Synthesis and Characterization of Polyurethane Nanoclay Composites. *International Journal of Polymer Science*, Article ID 717895, 5 pages.

[17] Barick A, Tripathy D. 2010 Effect of Organoclay on the Morphology, Mechanical, Thermal, and Rheological Properties of Organophilic Montmorillonite Nanoclay Based Thermoplastic Polyurethane Nanocomposites Prepared by Melt Blending. *Polymer Engineering and Science*, 484-498.

IOP Conf. Series: Materials Science and Engineering **195** (2017) 012002 doi:10.1088/1757-899X/195/1/012002

The estimation of quantitative parameters of oligonucleotides immobilization on mica surface

T I Sharipov[1], R Z Bakhtizin

Department of Physical Electronics and Nanophysics,
Bashkir State University, 450076 Ufa, Russia

E-mail: sha-T@yandex.ru

Abstract. Immobilization of nucleic acids on the surface of various materials is increasingly being used in research and some practical applications. Currently, the DNA chip technology is rapidly developing. The basis of the immobilization process can be both physical adsorption and chemisorption. A useful way to control the immobilization of nucleic acids on a surface is to use atomic force microscopy. It allows you to investigate the topography of the surface by its direct imaging with high resolution. Usually, to fix the DNA on the surface of mica are used cations which mediate the interaction between the mica surface and the DNA molecules. In our work we have developed a method for estimation of quantitative parameter of immobilization of oligonucleotides is their degree of aggregation depending on the fixation conditions on the surface of mica. The results on study of aggregation of oligonucleotides immobilized on mica surface will be presented. The single oligonucleotides molecules have been imaged clearly, whereas their surface areas have been calculated and calibration curve has been plotted.

1. Introduction

Immobilization of nucleic acids on the surface of various materials is increasingly being used in research and some practical applications. Currently, the DNA chip technology is rapidly developing. To create DNA chip a major component is attachment the DNA or DNA fragments - oligonucleotides - on the surface [1-2]. The basis of the immobilization process can be either physical adsorption (non-covalent fixation) [3-6] or chemisorption (covalent fixation) [7-10].

A proper way to control the immobilization of nucleic acids on a surface is to use atomic force microscopy (AFM) which allows one to investigate the topography of the surface by its direct imaging with high resolution [11-15]. In the review [8] the most common strategies for immobilization of biomolecules before the study by AFM have been represented. To fix the DNA on the surface of mica are used polyvalent cations which mediate the interaction between the negatively charged mica surface and the DNA molecules. Primarily, these transition metal ions of the fourth group. However, as shown in [10], it is possible to use monovalent metal cations.

Specific achieve visualization of short DNA fragments are shown for the first time in [13]. The authors showed that by using the AFM could visualize the DNA molecule length of about 25 nucleotides. Such single-stranded DNA is globule as capable of intramolecular interactions bases, especially in the presence of divalent metal cations.

The aim of the present work is to develop a method of estimation the quantitative parameter of immobilization of oligonucleotides: the degree of aggregation, depending upon the fixation conditions on the surface of mica.

IOP Conf. Series: Materials Science and Engineering **195** (2017) 012002 doi:10.1088/1757-899X/195/1/012002

2. Materials and Methods

We have used atomic force microscope Solver P47 and probe nanolaboratory Ntegra-Prima («NT-MDT», Zelenograd, Russia) for the study of oligonucleotides immobilized separately each one or combined into aggregates. Investigations were carried out in air using DLC (diamond-like) cantilevers («NT-MDT»). The substrate was fresh cleaved mica surface modified through treatment with metal cations. In the preparation of all solutions used deionized distilled water (Millipore, France). Specimens were 93-mer oligonucleotide ("Syntol", Moscow).

3. Results and discussion

First, it was verified reproducibility AFM images, i.e. the statistical analysis was performed. 10 samples were prepared by the same procedure. Immobilization solution containing 2 ng/μl DNA and 50 mM $MnCl_2$ was heated at 80° C for 2 minutes to denature the double-stranded structures. Then this solution was placed on the surface of fresh cleaved mica. The solution was kept for 4 min on the mica, and then the substrate was washed with water (100 μl). Each sample (substrate) in ten points scanned and respectively obtained 10 AFM images for each sample. On the obtained AFM images individual objects of different sizes were visualized. These probably as a single molecule of oligonucleotide and their aggregates. Heights of alleged single oligonucleotides were found, and calculated the arithmetic mean values of heights. According to the data the average height of objects situated on the surface of mica was found to be 0.58 ± 0.12 nm.

As an additional criterion for estimate the reproducibility was taken density of single oligonucleotides on the substrate. It was found that the density is 13 ± 2 pcs/μm^2. This indicates that the reproducibility of the AFM images is satisfactory and such criteria can be recommended to estimate the reproducibility of the obtained AFM images of biomolecules.

When immobilization of oligonucleotide molecules from solution they are arranged on the surface either in a single state or form aggregates. To estimate the number of molecules in the aggregate was the assumption that the surface area of the aggregate S_a should be independent of the number of molecules in aggregate n. In the same method were prepared the samples of different length oligonucleotides and then were visualized by AFM. Next, find a supposedly single oligonucleotides and their surface area was calculated as follows. On the AFM image a single oligonucleotides - a circular form objects - modeled the regular cone height h, base radius R and the generatrix L. Lateral surface of the cone $S_б = \pi RL = \pi R\sqrt{(h^2 + R^2)}$. Then knowing h and R can be calculated $S_б$. On the AFM image height and radius of the quasi-round form objects can be calculated by plotting their profiles, which are easily obtained using the integrated package of image processing software Image Analysis. Thereby calculating the surface area of the cones simulating single oligonucleotide molecules of different lengths on the AFM images, the calibration curve was plotted (Figure 1). From this calibration curve can be determined aggregate weight, and accordingly, the number of molecules in aggregate.

IOP Conf. Series: Materials Science and Engineering **195** (2017) 012002 doi:10.1088/1757-899X/195/1/012002

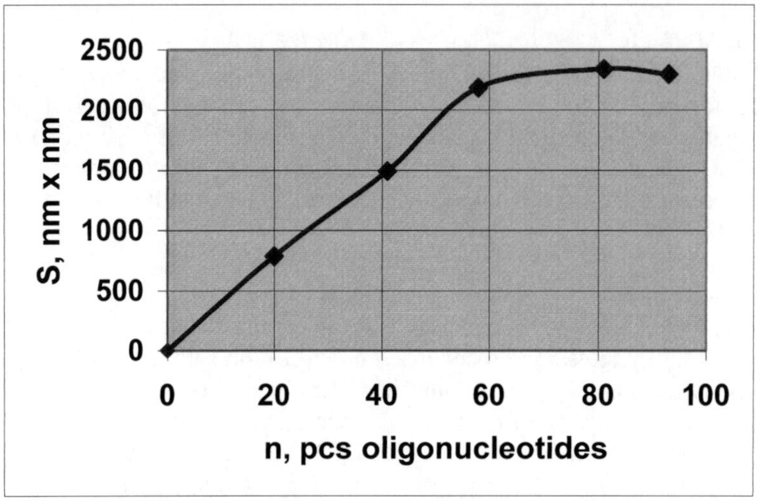

Fig. 1. Calibration curve for determining the number of oligonucleotide molecules in the aggregate

Figure 2 show example of determination of the number of molecules in the aggregate.

h, nm	R, nm	S_b, nm^2
1	27	2290.629

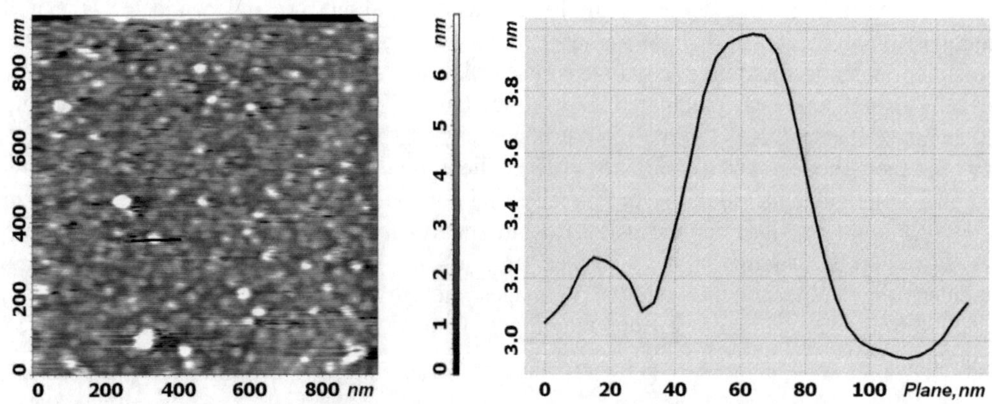

a b

Fig. 2. AFM image of oligonucleotides (a) and corresponding section profile (b).

Figure 2a show the AFM image of oligonucleotides (20 nucleotides length) immobilized on the mica surface. Figure 2b show the section profile obtained along the section line drawn in the figure 2a; in addition here are shown size and lateral area of the selected object. Knowing the area, a calibration curve is determined the size of the aggregate. For an object (Figure 2b), it is rounded is 80 nucleotides, i.e. this aggregate consists of four oligonucleotide molecules length of 20 nucleotides.

In order to find the dependence of the density of single oligonucleotides on the concentration of the immobilization solution components we performed two set of experiments. First, were study the dependence of density of single oligonucleotides ρ on their concentration in the immobilization solution (Figure 3) at a constant ratio of the number of oligonucleotide phosphate groups and cations Mn^{2+} 1:1. Studied objects were oligonucleotides length of 20 nucleotides.

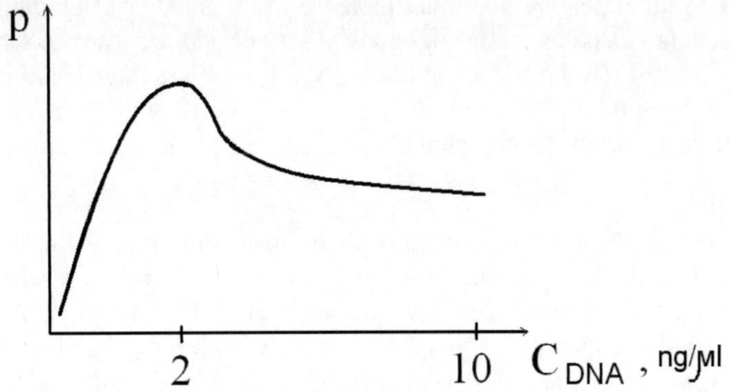

Fig. 3. The dependence of the density of single oligonucleotide molecules ρ on their concentration in the immobilization solution.

It was found that increasing the concentration of oligonucleotide molecules C_{DNA} from 0 to 2 ng/μl, there is a sharp increase ρ. In oligonucleotide concentration equal to 2 ng/μl, the density of single molecules is maximized. Then with increasing concentrations over 2 to 10 ng/μl observed a smooth curve with a gradual decline in the establishment of certain fixed amount per single immobilized molecules. Probably the increasing character of the curve due to the fact that at low concentrations of oligonucleotide molecules are not aggregated and immobilized are separately from each other. A gradual increase in the concentration of oligonucleotides increases the density of single molecules and aggregates. When the concentration of oligonucleotides more than 2 ng/μl aggregation process begins to dominate over the single immobilization process.

There was also investigated the dependence of the density of single oligonucleotide molecules ρ on the Mn^{2+} cations concentration in the immobilization solution $C_{Mn}{}^{2+}$. This dependence can be described by a function $\rho=f(C_{DNA}/C_{MnCl2})$ (Figure 4).

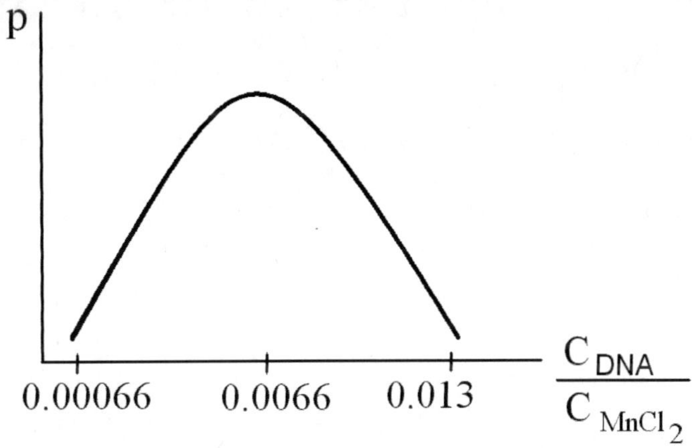

Fig. 4. The dependence of the density of single oligonucleotide molecules immobilized on the mica surface on the Mn^{2+} cations concentration.

When the concentration of Mn^{2+} cations in the immobilization solution increase from 25 to 50 μM (i.e. with value C_{DNA}/C_{MnCl2} from 0.013 to 0.0066) was observed increase in the density of single oligonucleotide molecules that may be due to an increase in the number of binding sites between the negatively charged oligonucleotide molecules and the surface of mica. When Mn^{2+} cations concentration above 50 μM (when the value C_{DNA}/C_{MnCl2} is less than 0.0066) oligonucleotide aggregation process in the solution increases so that the density of single oligonucleotide molecules immobilized on the surface is significantly reduced.

4. Conclusion

We investigate the reproducibility of the AFM images by measuring the height and density of single oligonucleotides immobilized on a substrate. We propose a method for plotting the calibration curve. This calibration curve illustrate dependence of the surface area of the cone, simulating oligonucleotide molecule, on the molecule length and thus the mass. Examples of determining the number of molecules in the aggregate are given. The dependence of the density of single oligonucleotides on the obtained AFM images on the concentration of the immobilization solution components has been studied.

The results of study may contribute to understanding the mechanisms of physical adsorption of DNA molecules and their fragments of different lengths on the surface of mica and may used for the design of DNA chip.

5. References

1. Lemieux B, Aharoni A and Schena M 1998 *Molecular Breeding* **4** 277–89
2. Gabig M and Wêgrzyn G 2001 *Acta Biochimica Polonica* **48 № 3** 615-22
3. Higashi N, Inoue T and Niwa M 1997 *Chem. Commun* **33** 1507-08
4. Müller D J, Engel A and Amrein M 1997 *Biosensors&Bioelectronics* **12 № 8** 867–77
5. Tanigawa M and Okada T 1998 *Analytica Chimica Acta* **365** 19-25
6. Lyubchenko U and Shlyakhtenko L 1997 *Proc. Natl. Acad. Sci.* **94** 496–501
7. Mirzabekov A D, Prokopenko D V and Chechetkin V R 2002 *In.: Biomedical Information Technology (edited by Knyazheva V A and Sudakov K V)* Moscow: GEOTAR-MED pp 166-98
8. Wagner P. 1998 *FEBS Letters* **430** 112-115
9. El Kirat K, Burton I, Dupres V and Dufrene Y 2005 *Journal of Microscopy* **218** 199–207
10. Hu J, Wang M, Weier H-U G, Frantz P, Kolbe W, Ogletree D F and Salmeron M 1996 *Langmuir* **12. №7** 1697-1700
11. Vesenka J, Guthod M, Tang C L, Keller R, Delaine E and Bustamante C 1992 *Ultramicroscopy* **42-44** 1243-49
12. Hansma H et al. 1992 *Nucleic Acids Research* **20** 3585-90
13. Hansma H, Revenko I, Kim K and Laney D 1996 *Nucleic Acids Research* **24. №4** 713-20
14. Mironov V L 2009 *Osnovy skaniruuschei zondovoi mikroskopii*. (Moscow: Tekhnosfera)
15. Ellis J S, Abdelhady H G, Allen S, Davies M C, Roberts C J, Tendler S J B and Williams P M 2004 *Journal of Microscopy* **215** 297–301

Acknowledgments

This work was supported by the Russian Foundation for Basic Research under grant No. 14-02-97022.

Feasibility of Plasma Treated Clay in Clay/Polymer Nanocomposites Powders for use Laser Sintering (LS)

Alaa Almansoori[1,3] , Ryan Seabright[1], C Majewski[2], and C Rodenburg[1]
[1] Material Science and Engineering Department, University of Sheffield
[2] Mechanical Engineering Department, University of Sheffield

E-mail: university email ammalmansoori1@sheffield.ac.uk
personal email: (ala_tech@yahoo.com)

Abstract. The addition of small quantities of nano-clay to nylon is known to improve mechanical properties of the resulting nano-composite. However, achieving a uniform dispersion and distribution of the clay within the base polymer can prove difficult. A demonstration of the fabrication and characterization of plasma-treated organoclay/Nylon12 nanocomposite was carried out with the aim of achieving better dispersion of clay platelets on the Nylon12 particle surface. Air-plasma etching was used to enhance the compatibility between clays and polymers to ensure a uniform clay dispersion in composite powders. Downward heat sintering (DHS) in a hot press is used to process neat and composite powders into tensile and XRD specimens. Morphological studies using Low Voltage Scanning Electron Microscopy (LV-SEM) were undertaken to characterize the fracture surfaces and clay dispersion in powders and final composite specimens. Thermogravimetric analysis (TGA) testing performed that the etched clay (EC) is more stable than the nonetched clay (NEC), even at higher temperatures. The influence of the clay ratio and the clay plasma treatment process on the mechanical properties of the nano-composites was studied by tensile testing. The composite fabricated from (3% EC/N12) powder showed ~19 % improvement in elastic modulus while the composite made from (3% NEC/N12) powder was improved by only 14%). Most notably however is that the variation between tests is strongly reduced when etch clay is used in the composite. We attribute this to a more uniform distribution and better dispersion of the plasma treated clay within polymer powders and ultimately the composite.

1. Introduction

The demand for materials that can solve weight and cost problems of metals in automobile and aerospace industry have raised interest in polymer-matrix nanocomposites. Unlike metals, polymers have desirable properties such as being easy to manufacture, light weight, and low cost, but their mechanical properties are limited compared to metals. To enhance the tensile strength and modulus of the polymers, reinforcements are incorporated in polymers in a low concentration ratio [1]. Nanoclay is one of the most interesting materials used as a reinforcement to enhance the strength and modulus of polymers [2]. However, the differences in degree of polarity and incompatibility between polymers and clays are limitations that restrict the strength of reinforcement. Hence, surface modification is used to render nanoclay miscible with polymer [3]. Since the early nineties, many studies have been conducted in polymer nanocomposites based on organically treated nanoclays.

[3] To whom any correspondence should be addressed.

There have been few studies on the treatment of clay using the plasma processes [4, 5] and even fewer attempts have been made to prepare polymers nanocomposites based on plasma treated nanoclay [6, 7]. Polymer nanocomposites can be produced in either conventional (melt compounding) or non-conventional manufacturing (e.g. laser sintering) processes [8]. The weak interaction between clays and matrices and the agglomeration of clay particles are the main challenges for all methods. To a certain extent, organo-treated nanoclays have increased the mechanical properties of polymers in the conventional manufacturing methods such as melt compounding [9]. However, when laser sintering was used, the incorporation of Cloisite 30B (organically modified nanoclay) in Nylon12 in the laser sintering technique has reduced the strength and the elongation at break [11], see table 1.
Hence, the conventional modification of nanoclay for use in the laser sintering is not enough to render it compatible with polymers. Thus, this study aimed to investigate the effect of the air plasma treatment technique on the surface compatibility of the nanoclays with polymers. Downward heat sintering is used in this study to mirror the laser sintering process.

Table 1. Summary of results from previous studies on nylon 12 and nylon12/nanoclay composites (brackets relative change with regards to neat N12)

	Phang et al (2005) [9]	Yan et al (2011) [10]	Jain et al (2010) [11][b]	Current study EC-based composite
Clay %		**Elastic Modulus, MPa**		
0%	1045±51.6	1420±250	736	853±28
1%	1110±25.8(6.2±5.5)%	xxx	xxx	xxx
2%	1140±77.4(9.1±8.9)%	xxx	xxx	xxx
3%	xxx[a]	1870±240(31.7±25)%	808.7(9.9%)	1011.7±10(18.6±3.5)%
5%	1480±100(41.6±10.9)%	xxx	755.7(2.7%)	998±17(17±3.9)%
Clay %		**Tensile strength, MPa**		
0%	33.56±0.18	38.3±2	49.15	39.58±0.68
1%	34.94±0.18(4.1±0.75)%	xxx	xxx	xxx
2%	34.75±0.18(3.5±0.75)%	xxx	xxx	xxx
3%	xxx	47.2±1.9(23.2±7.3)%	48.42(-1.5%)	42.95±0.41(8.5±2)%
5%	34.63±0.25(3.2±0.91)%	xxx	36.5(-25.7%)	40.3±0.35(1.8±1.9)%
Clay %		**Elongation at break, %**		
0%	xxx	20.8±2.8	27.82	24.17±1.27
1%	xxx	xxx	xxx	xxx
2%	xxx	xxx	xxx	xxx
3%	xxx	17.7±3(-15±19.8)%	14.62(-47.4%)	18.44±2(-23.7±9.9)%
5%	xxx	xxx	10.37(-62.7%)	9.17±1.3(-62.1±8.3)%

[a] xxx means no data is available
[b] Only absolute values are available in [11]

2. Materials

Polyamide 12 or Nylon 12 (trade name: Duraform PA2200; supplier: e-Manufacturing Solutions (EOS)) was used in this study as a matrix. Nylon 12 powder received had been used at least twice before in the laser sintering machines in the laboratories of the University of Sheffield. The chemical formula of Nylon12 is $(C_{12}H_{23}NO)_n$. The nanoclay used is organomodified montmorillonite (trade name: Cloisite 30B (supplier: Southern Clay Products) described further in [12].

3. Preparation Methods

3.1 Plasma treatment technique
Air/Oxygen Plasma treatment (also known as Plasma Etching) was used to treat C30B for ~30 minutes before adding the nanoclay to nylon 12. Plasma Cleaner Zepto from Diener Electronic is available in the Sorby Centre at the University of Sheffield. The nanoclay powder placed in glass petri dishes inside the chamber of the plasma cleaner was subjected twice to 100 % power and 2-4 mbar for 1000 sec. In the current study, the plasma treated clay is called as etched clay (EC) and the other clay is called as nonetched clay (NEC).

3.2 Dry Mixing
In atmospheric environment, 0.15, 0.25, and 0.35 gram of EC and NEC powders were added to 5 g of Nylon 12 powder in small glass jars. After that, the mixed powder was stirred in a magnetic stirrer for a half hour. Then the composite was sonicated in an ultrasonicator bath (model FB15051 made by Fisher company) for also half. Composition are summarized in table 2.

Table 2. The concentration of nanoclay (EC and NEC) in the nylon12 nanocomposite.

EC or NEC	nanoclay, g	N12, g	Comp.
NEC	0.15	5	3%NEC/N12
NEC	0.25	5	5%NEC/N12
EC	0.15	5	3%EC/N12
EC	0.25	5	5%EC/N12

3.3 Downward Heat Sintering
Downward heat sintering (DHS) in a hot press is used to process neat and composite powders into tensile and XRD specimens in order to reflect laser sintering conditions. Powders were placed in-between the two parts of the hot press. First, the powders is left in a hollow mould on the lower part (LP) for 15 minutes where the temperature of LP is just below the melting temperature (LP temperature is 185 °C for the neat N12 and 188 °C for N12 composites). Thus, the powder is preheated before the upper part (UP) goes down to generate a close heat chamber for another 15 minutes (UP temperature is 190 °C for the neat N12 and 193 °C for N12 composites). The temperature of the powder, is then raised to reach value above the melting temperature of N12. After that the powder being fully melted and the mould and samples are left to be cooled outside the hot press in at room temperature.

4. Characterization and Testing Techniques

4.1 X-Ray Diffraction (XRD)
XRD of powder was carried out using a Siemens D5000 (Cu, GAXRD) at room temperature from $2\theta = 2^{o}$ to 27° in a step of 0.02^{o} and 1 sec per step. The machine was operated at 40 KV and 40 mA.

4.2 Thermogravimetric Analysis (TGA)
A Thermogravimetric Analyser (Pyris 1 TGA from PerkinElmer) was used. All powders were heated, in Nitrogen atmosphere, from 30 °C to 630 °C with a heating rate of 10 °C/min.

4.3 Scanning Electron Microscopy (SEM)
A Nova NanoSEM (Low Voltage Scanning Electron Microscopy (LV-SEM)) was used.

4.4 Tensile Testing
Tensile test was carried out using a Hounsfield Tensometer according to BS ISO 527. A 10000 N load cell, speed of test was 5 mm/min and a preload of 5 N were used.

IOP Conf. Series: Materials Science and Engineering **195** (2017) 012003 doi:10.1088/1757-899X/195/1/012003

5. Results and Discussion

5.1 XRD results

XRD patterns are shown in figure 1 (a). NEC exhibited a well-defined diffraction peak at 2θ of 4.8° and the interlayer spacing is equal to 1.8 (001 crystal lattice). While, the X-Ray scattered from EC gave a multiple peak at the same position of NEC, and the multiple peak tends to scatter more diffuse and have a low intensity peak than one sharp peak. Both NEC and EC spectra exhibited a weak peak with the same intensity at 2θ of 19.7°, and this is similar to what is reported in reference [7].

5.2 TGA results

Figure 1 (b) shows weight loss as a function of temperature. Organo-modified nanoclay is degraded in fourth stages, i.e. desorption of water, dehydration of hydrated cation, loss of surfactant, and dihydroxylation [13]. According to the results in figure 1 (b), all those steps were shifted to higher temperature compared with NEC. Hence, the EC is more stable than the NEC, even at higher temperatures.

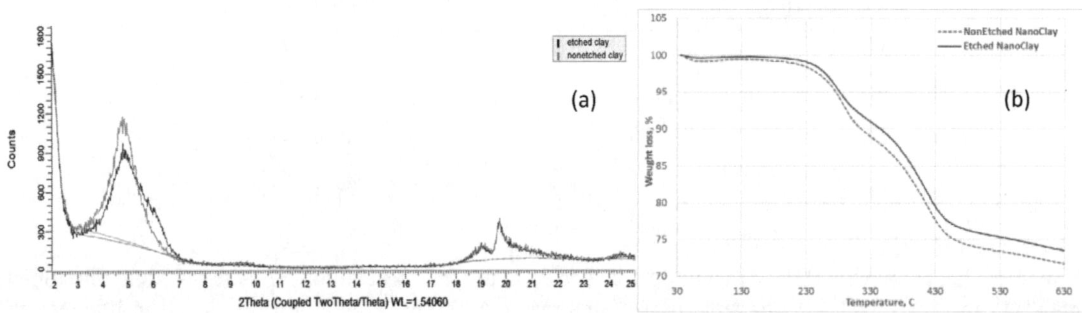

Figure 1. A comparison between the Etched and nonetched results using (a) XRD (b) TGA techniques

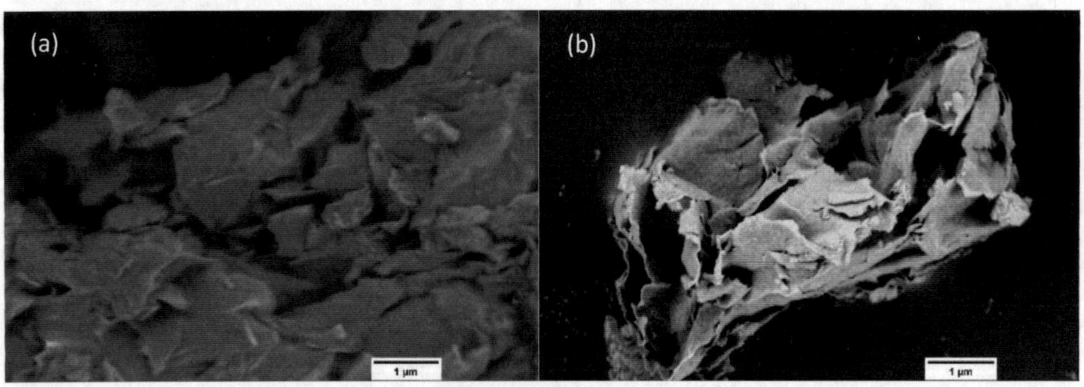

Figure 2. SEM images of (a) nonetched clay, and (b) etched clay

5.3 SEM images

The SEM images of NEC and 30 minutes treated EC are shown in figure 2 (a and b) respectively. Figure 2 (a) shows the clay platelets are stuck together which may result in non-uniform clay distribution. Non-uniform clay agglomerates on the surface of polymer particles and can weaken the strength of the composites [11]. Plasma treatment minimizes the sticking of clay platelets as shown in figure 2 (b).

5.4 Tensile test results

The tensile test results are presented in figures 3 and 4 (a). Adding the nanoclay (etched and nonetched) at concentrations 3% and 5% has improved the elastic modulus and tensile strength of N12 but leads to a reduction in the elongation at break. It is important to consider the combination of these properties [14]. The best combination of all three properties is found for the 3% EC/N12 nano-composite. While the elongation at break was reduced by ~24% compared to neat nylon 12, the elastic modulus is increased by ~19% with a simultaneous increase in tensile strength of ~9%. Both exceed the performance of clay/N12 laser sintered nanocomposites with the same clay loading reported in [11], see table 1. The reduction in the elongation is less pronounced for the EC/N12 nanocomposite (~24%) compared to ~52% for the NEC/N12 nanocomposite. For EC/N12, the fracture surface shows sign of ductility (see figure 4 (b). The main reason for the reduced elongation and the brittle fracture is the adding of rigid clay [10]. In addition, the interaction between the clay and polymer is not strong enough to resist the axial force. This leads to the generation of microvoids [9] which develop to initiate a micro crack and the propagated cracks will lead to a brittle fracture. Based on this mechanism our results imply that the interaction between clay and polymer in the EC/N12 nanocomposite is stronger than that in NEC/N12 nano-composites. Also notable is that the variation between tests is strongly reduced when etched clay is used in the composite (see figure 3). We attribute this to a more uniform distribution and better dispersion of the plasma treated clay within polymer powders and ultimately the composite.

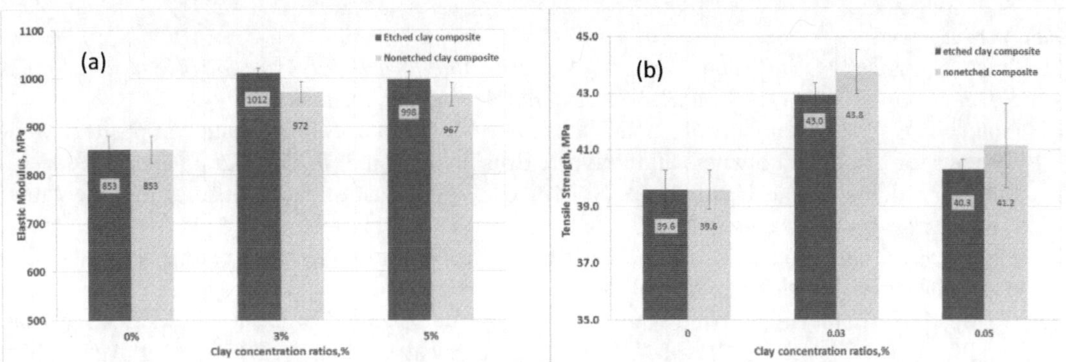

Figure 3. (a) Elastic modulus and (b) Tensile strength of EC and NEC based N12 composites.

At high filler loadings some of the clay platelets may not well disperse and that may negatively impact the tensile strength [11] and the elongation at break [9]. This seems to be the case for the 5% clay loading in our experiments, in which the elastic modulus is similar to materials with 3% clay but the tensile strength is hardly improved while the elongation at break was dramatically reduced by 62% (EC/N12) of the value of neat N12.

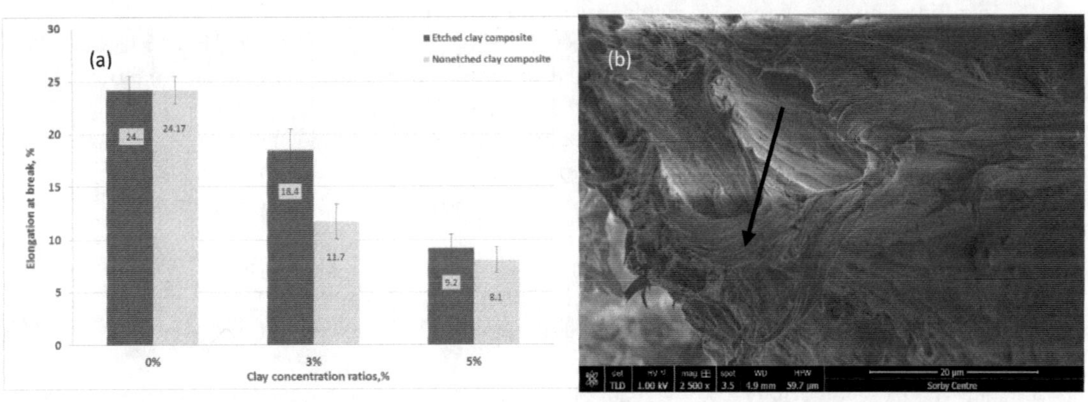

Figure 4. (a) The elongation at break values of N12 after adding etched and nonetched clay, (b) SEM image of 3%EC based nylon12 nanocomposite showing features of ductile fracture.

6. Conclusions

An improvement in the tensile properties (elastic modulus and tensile strength) when plasma treated organoclay is incorporated into non-virigin (powder used previously in laser sintering) nylon12 powder by a simple, dry mixing process is achieved. With this method, the tensile results after adding organoclay at concentrations of 3% or 5% are almost the same. However, in term of the cost of the material and the elongation of parts, the 3% based composite is the best. It means the method for making nanocomposite with enhanced properties at lower filler concentration is achieved.

The obtained polymer nanocomposites compare favourably to other mixing process and are encouraging for the use of dry mixed powders in Laser Sintering.

7. Acknowledgement

A. Almansoori would like to thank financial support from MOHESR of IRAQ. C. Rodenburg thanks EPSRC for funding under EP/N008065/1. We also thank Dr Biqiong Chen, Wendy M Birtwistle, and Robert Master and also the staff of Sorby centre for Electron Microscopy and Microanalysis, are greatly acknowledged for their support.

8. References

[1] Gupta N, Pinisetty D and Shunmugasamy V C 2013 *Reinforced Polymer Matrix Syntactic Foams: Effect of Nano and Micro-Scale Reinforcement* (New York: Springer).

[2] Giannelis E P, Krishnamoorti, R, and Manias E 1999 Polymer-Silicate Nanocomposites: Model Systems for Confined Polymers and Polymer Brushes *Advances in Polymer Science* **138**.

[3] Singla P, Mehta R, and Upadhyay S N 2012 Clay modification by the use of organic cations *Green and sustainable chemistry* **2** 21-25.

[4] Fatyeyeva K and Poncin-Epaillard F 2011 Sulfur dioxide Plasma Treatment of the Clay (Laponite) Particles *Plasma Chem Plasma Process* **31** 449–464.

[5] Djowe A T, Laminsi, S, Njopwouo, D, Acayanka, E and Gaigneaux E M 2013 Surface Modification of Smectite Clay Induced by Nonthermal Gliding Arc Plasma at Atmospheric Pressure *Plasma Chem Plasma Process* **33** 707–723.

[6] Shamsi M H, Luqman M, Basarir, F, Kim J S, Yoona T H and Geckelera K E 2010 Plasma-modified halloysite nanocomposites: effect of plasma modification on the structure and dynamic mechanical properties of halloysite–polystyrene nanocomposites *Polym Int* **59** 1492–1498.

[7] Scaffaro R, and Maio A, Enhancing the mechanical performance of polymer based Nanocomposites by plasma-modification of nanoparticles 2012 *Polymer Testing* **31** (2012) 889–894

[8] Athreya S R, Kalaitzidou, K, and Das S 2012 Microstructure Thermomechanical Properties and Electrical Conductivity of Carbon Black-Filled Nylon-12 Nanocomposites Prepared by Selective Laser Sintering, *polymer engineering and science* **52(1)** 12-20

[9] Phang I Y, Liu T, Mohamed A, Pramoda K P, Chen L, Shen L, Chow S Y, He C, Lu X and Hu X 2005 Morphology, thermal and mechanical properties of nylon 12/organoclay nanocomposites prepared by melt compounding *Polym Int.* **54** 456–464.

[10] Yan C Z, Shi Y S, Yang J S and Liu J H 2011 An organically modified montmorillonite/nylon-12 composite powder for selective laser sintering *Rapid Prototyping Journal* **17** **(1)** 28 – 36.

[11] Jain P K, Pandey P M, and Rao P V M 2010 Selective Laser Sintering of Clay-Reinforced Polyamide *Polymer Composites* **31** **(4)** 732–743.

[12] Lee K-S and Kubayashi S 2010 Polymer Materials: Block-Copolymers, Organic/Inorganic Hybrids, Polymethylenes (Verlag Berlin Heidelberg-Springer) page 9

[13] Singla P, Mehta R, and Upadhyay S N 2012 Clay Modification by the Use of Organic Cations *Green and Sustainable Chemistry* **2** 21-25

[14] Chen B and Evans J R G 2009 Impact strength of polymer-clay nanocomposites *Soft Matter* **5** 3572–3584.

IOP Conf. Series: Materials Science and Engineering **195** (2017) 012004 doi:10.1088/1757-899X/195/1/012004

Chitosan-Based Nanocomposite Beads for Drinking Water Production

ML Masheane[1,2], LN Nthunya[1,2], SS Sambaza[2], SP Malinga[2], EN Nxumalo[1], BB Mamba[1] and SD Mhlanga[1*]

[1]Nanotechnology and Water Sustainability Research Unit, College of Science, Engineering and Technology, University of South Africa, Florida, 1709, Johannesburg, South Africa.

[2]Department of Applied Chemistry and the DST/Mintek Nanotechnology Innovation Centre-Water Research Node, University of Johannesburg, P.O. Box 17011, Doornfontein, 2028, South Africa.

E-mail: mhlansd@unisa.ac.za

Abstract. Potable drinking water is essential for the good health of humans and it is a critical feedstock in a variety of industries such as food and pharmaceutical industries. For the first time, chitosan-alumina/functionalised multiwalled carbon nanotube (f-MWCNT) nanocomposite beads were developed and investigated for the reduction of various physico-chemical parameters from water samples collected from open wells used for drinking purposes by a rural community in South Africa. The water samples were analysed before and after the reduction of the identified contaminants by the nanocomposite beads. The nanocomposite beads were effective in the removal of nitrate, chromium and other physico-chemical parameters. Although, the water samples contained these contaminants within the WHO and SANS241 limits for no risk, the long-term exposure and accumulation is an environmental and health concern. The reduction of these contaminants was dependent on pH levels. At lower pH, the reduction was significantly higher, up to 99.2% (SPC), 91.0% (DOC), 92.2% (DO), 92.2% (turbidity), 96.5% (nitrate) and 97.7% (chromium). Generally, the chitosan-alumina/f-MWCNT nanocomposite beads offer a promising alternative material for reduction and removal of various physico-chemical parameters for production portable water.

Key words: Beads, carbon nanotubes, chitosan, nanocomposites, water purification

1. Introduction

The contamination of water resources by toxic molecules of organic, inorganic and biological nature is currently the world's major concern. Potable water production from surface and groundwater sources normally implicates the use of water treatment processes that remove turbidity, bacteria, colour, organic compounds and clay particles. These suspended materials are of environmental concern and pose a serious threat to water quality consumed by societies globally (Bina et al. 2009). Hence, these contaminants must be adequately removed from water in order to produce potable water and to comply with stringent environmental quality standards. Recently, there has been a significant interest in the development of non-toxic, biocompatible and biodegradable biopolymers such as chitosan, that can used in water treatment (No and Meyers 2000).

Chitosan is a natural linear biopolyaminosaccharide obtained by the alkaline deacetylation of chitin, which is the principal component of protective cuticles of crustaceans such as crabs, shrimps, prawns and lobsters. It is a weak base that is insoluble in water and in organic solvent but soluble in dilute

* To whom correspondence should be addressed - email: mhlansd@unisa.ac.za

aqueous acidic solutions (pH < 6.5), which can convert glucosamine units into soluble forming R–NH_3^+ (Dutta et al. 2004). Chitosan has recently emerged as a low-cost biosorbents for the removal of heavy metals, organics and micro-organisms. Although it is cost effective and non-toxic, it is unstable in harsh conditions such as low pH levels and higher temperatures. To overcome this challenge, scientists have focussed on its modification. The presence of three of the reactive groups on chitosan, an amino group at C-2 position, as well as the primary and secondary hydroxyl groups at C-6 and C-3 positions, respectively make this material very susceptible to chemical modification such as acetylation, N-phthaloylation, tosylation, alkylation and graft polymerization (Olteanu 2007). Modification of chitosan with inorganic materials is of research interest.

The objective of this work was to synthesize a novel adsorbent, chitosan-alumina/functionalised multiwalled carbon nanotube (f-MWCNT) nanocomposite beads for the removal of total chromium and nitrates from water. The physico-chemical parameters such as electrical conductivity, turbidity and total dissolved solutes were also studied to ascertain the validity of the nanocomposite beads for the removal of chromium and nitrates from water.

2. Experimental

2.1. Materials

Chitosan (degree of deacetylation= 85%), sodium nitrate, potassium dichromate, alumina (particle size in the range of 30-60 nm), sodium hydroxide, hydrochloric acid, oxalic acid, nitric acid and sulfuric acid were purchased from Sigma Aldrich. MWCNTs were synthesised and acid treated in our laboratory using the procedure by Mhlanga et al. 2009. Deionized water was obtained from a water purification system (Po process Ecopure G.I.I Scientific) available in our laboratories.

2.2. Preparation of chitosan-alumina/f-MWCNT nanocomposite beads

Chitosan gel (2%) was prepared using a modified procedure reported by Ngah and Fatinathan 2010, and Chatterjee and Woo 2009. Chitosan powder (2 g) was dissolved into 100 ml of 5% oxalic acid and stirred for 12 h. Then 0.2 g alumina and 0.1 g f-MWCNTs were added while stirring. The homogeneous gel was de-gassed and added dropwise into a precipitation bath containing 2 M sodium hydroxide solution by using a 0.4 mm cut diameter syringe needle and vigorously stirred using a magnetic stirrer. After a few seconds, black coloured beads formed, and they were collected and washed with distilled water until pH 7 was attained. The beads were dried in the oven at 60 °C overnight prior to use.

2.3. Batch adsorption experiments

The experiments were conducted in a 250 mL conical flask containing 100 mL of natural water samples using 1 g of chitosan and chitosan based nanocomposite beads. The flask and its contents were agitated at room temperature using a linear shaker. The effect of pH was studied at various pH levels ranging from pH 2 – 8 at the 1 g/100 mL beads dosage for 60 min contact time. Finally, the effect of contact time on the removal efficiency was investigated in the range of 5 – 240 min. After the experiment, the mixtures were then filtered and the concentrations of total chromium and nitrate were analysed using an inductively coupled plasma optical emission (ICP-OES) ICAP 6000 series spectrometer thermos Fisher-Scientic and Dionex Ion Chromatography 2000 (IC) spectroscopy respectively. The removal efficiency of nitrates, total chromium and various physico-chemical parameters was calculated using **Equation 1** (Jagtap et al. 2011):

$$\% \text{ Removal} = \frac{C_i - C_e}{C_i} \times 100 \quad (1)$$

where C_i and C_e are the initial and final concentrations (g/mL), respectively.

IOP Conf. Series: Materials Science and Engineering **195** (2017) 012004 doi:10.1088/1757-899X/195/1/012004

2.4. Sampling

The water samples were collected from a shallow well in rural community (Lochiel) in Mpumalanga province in South Africa. This well is used by the community for drinking and other domestic purposes. The water samples were collected in plastic containers which were sterilised with methanol, and carried in a cooler box to laboratory, where they were stored in the refrigerator at 4 °C. The samples were also analysed on-site for dissolved oxygen (DO), electrical conductivity, pH, temperature and alkalinity using Hanna Instrument multi-meter (model Hi-9828).

3. Results and discussion

3.1. Physical properties of chitosan-alumina/f-MWCNT nanocomposite beads

Figure 1 shows the chitosan-alumina/f-MWCNT nanocomposite beads. The nanocomposite beads were black in colour due to the presence of black functionalised MWCNTs. The nanocomposite beads were ellipsoidal in shape with an average diameter (size) in the range of 3 – 5 mm.

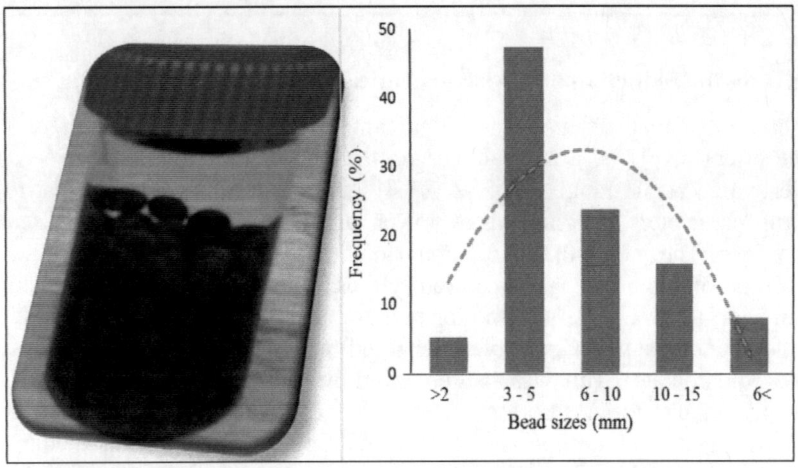

Figure 1: Chitosan-alumina/f-MWCNT nanocomposite beads. Insert is the average diameter (size) of the nanocomposite beads.

3.2. Reduction of physico-chemical properties by chitosan-based nanocomposite beads

The physical characteristics of water used by the community of Lochiel in Mpumalanga are presented in **Table 1**. Generally, the results of the analysis demonstrated that water samples collected from one of drinking water sources in Lochiel (shallow well) were within the WHO and the Department of Water Affair and Sanitation recommended limits for drinking water quality specifications for no risk in terms of pH (5 – 9), turbidity (< 5 NTU), DOC (< 8.2 mg/L), DO (< 1000 mg/L), chromium (< 100 µm/L), nitrate (< 6 mg/L) and SPC (< 150 µS/cm) (WHO 2011; SANS 2006). The reduction of these physico-chemical properties using chitosan- alumina/f-MWCNT nanocomposite beads was significantly high. The concentration of SPC, DOC, DO, nitrates and chromium drop from 26.8, 0.87, 8.71, 0.26 and 0.8 mg/L to 0.67, 0.10, 0.73, 0.01 and 0.02 mg/L, respectively. Turbidity also decreased from 4.62 to 0.05 NTU while pH was increased from 6.27 to 7.08, mainly because metals act as Lewis acids and their removal decrease acidity. Therefore, it is obvious that parameters were reduced largely by chitosan-alumina/f-MWCNT nanocomposite beads.

Table 1: Physico-chemical characteristics of raw and treated water from Lochiel.

Parameters	Raw Water	Treated water	%Removal
SPC (μS/cm)	26.84	0.67	97.51
pH	6.27	7.08	N/A
DOC (mg/L)	0.87	0.10	88.28
DO (%)	8.71	0.73	91.58
Turbidity (NTU)	4.62	0.05	98.97
Nitrates (mg/L)	0.26	0.00	98.17
Chromium (mg/L)	0.80	0.01	97.72

3.3. Effect of pH on the reduction of physico-chemical properties

pH is an important parameter that has a significant effect on the adsorption of species on the adsorbents. The effect of pH of water samples on the chitosan-alumina/f-MWCNT nanocomposite beads was investigated at pH range of pH 2 – 8 (Figure 2). It was observed that the reduction of physico-chemical parameters was higher at lower pH levels, and was decreased slightly with increasing pH values. The increase in the amount of nitrates adsorbed at low pH was attributed to protonation of some amino ($-NH_2$) groups from chitosan and hydroxyl ($-OH^-$) groups from either chitosan, alumina or f-MWCNTs. The protonation of the active sites significantly enhanced the electrostatic interactions between positively charged adsorbent surface and negatively charged sorbates (nitrates, dichromate). This was also observed by Demiral and Gündüzoğlu 2010 on positive impact of low pH on nitrates adsorption by activated carbon. **Equations 2 and 3** represent the protonation of $-NH_2$ and $-OH^-$ groups and removal of nitrates from aqueous media. The lower pH values cause the surface hydroxyl ($-OH^-$) groups to accept proton (H^+), thus facilitating the ligand exchange since water (H_2O) is a weak ligand than -OH to get displaced from adsorbent bonding sites. Basically, the lower pH levels promote the adsorption of anions, *i.e.* at low pH, a large number of hydrogen ions (H^+) can neutralise the oppositely charged surface and thus enhance the electrostatic attraction between the adsorbent and adsorbate, overcoming electrostatic repulsion between them (Alemayehu et al. 2012). Furthermore, at lower pH, the spontaneous reduction of chromium (VI) to chromium (III) was because of reduction-oxidation potential (13.4 mV), thus resulting to up to 97.7% chromium removal (Nithya and Sudha 2017).

$$-NH_2 + H^+ \rightarrow -NH_3^+ + NO_3^- \rightarrow -NH_3^+ \cdots\cdots NO_3^- \quad (2)$$

$$-OH^- + 3H^+ \rightarrow -OH_3^+ + NO_3^- \rightarrow -OH_3^+ \cdots\cdots NO_3^- \quad (3)$$

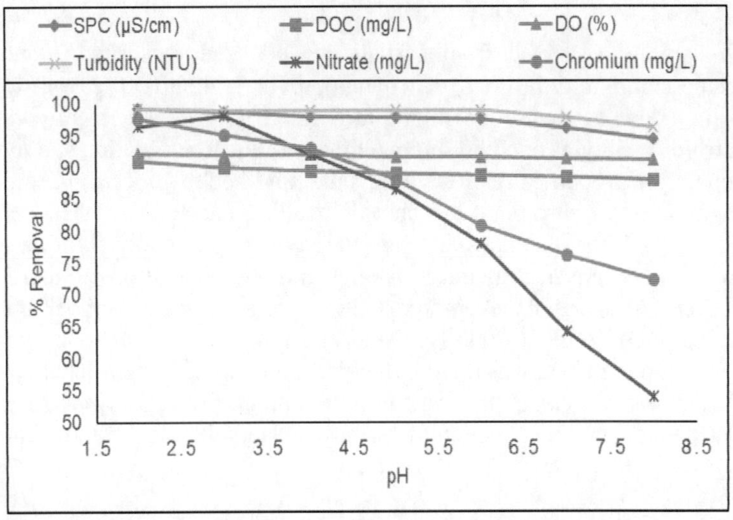

Figure 2: The effect of pH on the reduction of physico-chemical pamarameter using 1 g of chitosan-alumina/f-MWCNT nanocomposite beads at constant contact time (60 min).

3.4. Effect of contact time on the reduction of physico-chemical parameters

The effect of contact time on the reduction of physico-chemical properties by chitosan-alumina/f-MWCNT nanocomposite beads was investigated at 0 - 240 min time interval at pH 6 as shown in **Figure 3**. It was observed that there was a rapid reduction of physico-chemical parameters within the first 30 min and equilibrium water reached within 60 min. After 60 min, there is no significant change in the reduction of the physico-chemical reduction, meaning that the adsorbent active sites were being saturated. The highest reduction efficiencies for physico-chemical parameters achieved were 99.12% (SPC), 90.22% (DOC), 91.90% (DO), 99.85% (turbidity), 98.01% (nitrate) and 82.03% (chromium).

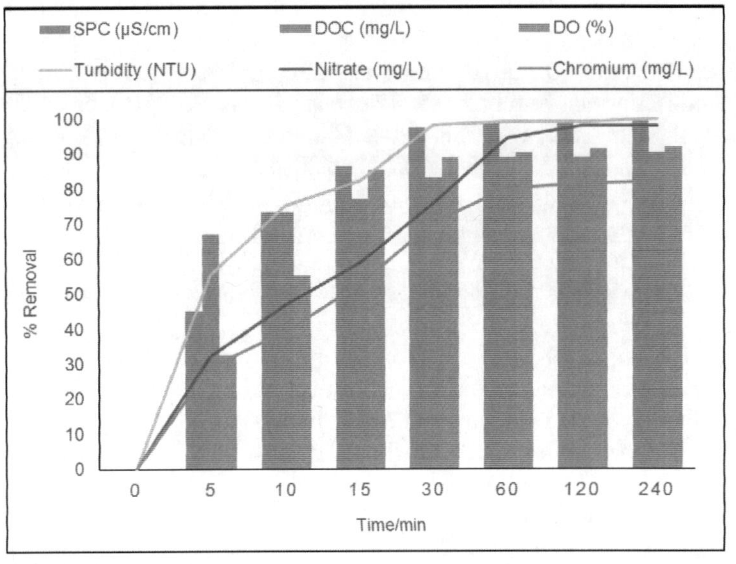

Figure 3: The effect of contact time on the removal of physico-chemical parameters by 1 g of chitosan-alumina/f-MWCNT nanocomposite beads.

4. Conclusions

The quality water samples collected from a shallow well used for drinking and other domestic purposes in a selected community in the Mpumalanga province, South Africa was found to be within WHO and SANS limit for no risk from drinking in terms of physico-chemical parameters including SPC, DO, DOC, turbidity, nitrates and chromium. However, the accumulation and long-term exposure of these physico-chemical properties is of environmental and health concern. For instance, long-term exposure and accumulation of chromium has chronic health effect (cancer). The newly development chitosan-alumina/f-MWCNT nanocomposite beads were effective in removal nitrate and chromium as well as reduction of other physico-chemical parameters. The reduction of these parameters was dependent on pH levels. At lower pH level the reduction was significantly higher, up to 99.2% (SPC), 91.0% (DOC), 92.2% (DO), 92.2% (turbidity), 96.5% (nitrate) and 97.7 % (chromium). The removal of the physico-chemical parameters was rapid in the first 30 min and insignificant beyond 60 min, due to saturation of the active sites. Generally, the chitosan-alumina/f-MWCNT nanocomposite beads can be employed in water treatment for reduction of various physico-chemical parameters.

Acknowledgements

The authors wish to thank (DST)/NRF Nanotechnology Flagship project (No 88046), the Water Research Commission, DST/Mintek Nanotechnology Innovation Centre (NIC), the Department of Applied Chemistry (University of Johannesburg), and the Nanotechnology and Water Sustainability (NanoWS) Research Unit for supporting this work. Part of this work was extraced from the masters dissertation of the first author, ML Masheane.

References

Alemayehu, D.D., Singh, S.K. and Tessema, D.A., 2012. Assessment of the adsorption capacities of fired clay soils from Jimma (Ethiopia) for the removal of Cr (VI) from aqueous solution. *Universal J. Environ. Res. Technol.*, 2(5), pp.411–420.

Bina, B., Mehdinejad, M.H., Nikaeen, M. and Attar, H.M., 2009. Effectiveness of chitosan as natural coagulant aid in treating turbid waters. *Iran. J. Environ. Health. Sci. Eng.*, 6(4), pp.247–252.

Chatterjee, S. and Woo, S.H., 2009. The removal of nitrate from aqueous solutions by chitosan hydrogel beads. *J. Hazard. Mater.*, 164(2-3), pp.1012–1018.

Demiral, H. and Gündüzoğlu, G., 2010. Removal of nitrate from aqueous solutions by activated carbon prepared from sugar beet bagasse. *Bioresour. Technol.*, 101(6), pp.1675–1680.

Dutta, P.K., Dutta, J. and Tripathi, V.S., 2004. Chitin and chitosan: Chemistry, properties and applications. *J. Sci. Industr. Res.*, 63(January), pp.20–31.

Jagtap, S., Yenkie, M.K.N., Labhsetwar, N. and Rayalu, S., 2011. Defluoridation of drinking water using chitosan based mesoporous alumina. *Micropor. Mesopor. Mater.*, 142(2-3), pp.454–463.

Mhlanga, S.D., Mondal, K., Carter, R., Witcomb, M.J. and Coville, N.J., 2009. The effect of synthesis parameters on the catalytic synthesis of multiwalled carbon nanotubes using Fe-Co/$CaCO_3$ catalysts. *S. Afr. J. Chem.*, 62, pp.67–76.

Ngah, W.S.W. and Fatinathan, S., 2010. Pb(II) biosorption using chitosan and chitosan derivatives beads: Equilibrium, ion exchange and mechanism studies. *J. Environ. Sci.*, 22(3), pp.338–346.

Nithya, R. and Sudha, P.N., 2016. Removal of heavy metals from tannery effluent using chitosan-g-poly(butyl acrylate)/bentonite nanocomposite as an adsorbent. *Textil. Cloth. Sustainability*, 2(1), p.1-7.

No, H.K. and Meyers, S.P., 2000. Application of chitosan for treatment of wastewaters. *Rev. Environ. Contamin. Toxicol.*, 163(1976), pp.1–27.

Olteanu, C.E., 2007. Applications of functionalized chitosan. *Sci. Stud. Res.*, 3, pp.227–256.

SANS-241, 2006. South African National Standard: Drinking water-drinking water. Published by Standards South African, a division of SABS. ISBN 0-626-17752-9.

WHO, 2011. World Health Organization. Guideline for drinking water quality. 20 Avenue Appia, 1211 Geneva 27, Swirtzerland.

IOP Conf. Series: Materials Science and Engineering **195** (2017) 012005 doi:10.1088/1757-899X/195/1/012005

Asphaltenes as new objects for nanoelectronics

M Yu Dolomatov[1,2], A M Petrov[1], R Z Bakhtizin[1], M M Dolomatova[1], I R Khairudinov[2], S A Shutkova[1], E A Kovaleva[1], N Kh Paymurzina[1]
[1] Department of physical electronics and nanophysics, Bashkir State University, 32, Validy Str., Ufa 450076, Russia
[2] Department of Technology of oil and gas, Ufa State Petroleum Technical University, 1, Kosmonavtov Str., Ufa 450062, Russia

E-mail: dolomatov@gmail.com

Abstract. Abstract. Modern carbon nanomaterials (carbon nanotubes, graphenes, fullerenes, polycyclic molecules) are products of rather complicated technologies. Therefore development of new not expensive materials on the basis of natural substances, in particular high-molecular compounds of oil – asphaltenes, is actual for nanoelectronics. Asphaltenes are complex materials that are found in crude oil, bitumen and high-boiling hydrocarbons distillates. Usually asphaltenes are composed mainly of polyaromatic carbon with a small amount of vanadium and nickel, which are in porphyrin structures. Molecules of asphaltenes may contain 5-10-member benzene and naphthenic rings in their structure and also have paramagnetic centers. A variety of techniques: electronic phenomenological spectroscopy (EPS), atomic force microscopy (AFM) and quantum chemistry calculations were used to define the structure of oil asphaltenes. It was supposed that asphaltene fraction is a strong donor (ionization potential 4.10-6.70 eV) and an acceptor (electron affinity 1.80-2.50 eV). The structures of asphaltenes fragments were calculated by RHF-6-31G** methods. AFM images of asphaltenes obtained from crude oil showed the presence of structure fragments ranged from 3 to 10 nm, disposed to strong intermolecular interactions. We used doped compounds for formation of wide band gap amorphous semiconductors from a concentrates of asphaltens. Changes of conductivity in dispersed petroleum systems (DPS) were studied during a pyrolysis at 500 K. The numerous experiments defined of conductivity testify about phase transitions dielectric – semiconductor in DPS for range of 360 – 400 K. The main conclusion is paramagnetic phase of asphaltenes is organic amorphous wide band gap semiconductor. Besides this substance can be consider as an organic spin glasses.

1. Introduction
Modern carbon nanomaterials are expensive products of difficult technology (carbon nanotubes, graphenes, fullerenes, polycyclic molecules). Therefore, search of new cheap materials based on natural substances, in particular high-molecular compounds of petroleum – asphaltenes, is actual for nanoelectronics. Asphaltenes are complex substances that are found in crude oil, bitumen and high-boiling hydrocarbons distillates [1]. Asphaltenes are composed mainly of polyaromatic carbon with vanadium and nickel traces, which are in porphyrin structures. Molecules of asphaltenes may contain 5-10-member benzene and naphthenic rings in their structure. Asphaltenes have paramagnetic centers. We had the research task to define electron, molecular and supramolecular structure of various asphaltenes.

2. Objects and methods
Experimental methods of electronic phenomenological spectroscopy (EPS) [2], atomic force microscopy (AFM) and quantum chemistry were used to definite the structure of petroleum asphaltenes. The EPS is based on the regularities of the relations between the absorption coefficients and electronic properties of the substance. Effective Ionization potential (EIP) and effective electron affinity (EEA) of

asphaltenes have been estimated according to empirical dependences linking these characteristics with integral force of oscillator (1) and (2) in UV and/or VIS spectrum regions under consideration:

$$E = \alpha_1 + \alpha_2\theta \quad (1)$$

$$\theta = \int_{\varepsilon_2}^{\varepsilon_1}\int_{\lambda_2}^{\lambda_1} f(\varepsilon_\lambda)d\varepsilon d\lambda \quad (2)$$

where E is effective ionization potential or effective electron affinity, eV; α_1, α_2 are empirically determined coefficients, eV and eV·mole·l^{-1}·nm^{-1} respectively; ε_λ is molar absorption coefficient; λ_1, λ_2 are borders of wavelengths in the spectrum in UV and (or) visible regions.

The structures of asphaltenes model fragments were calculated by RHF-6-31G** methods [3].

Electrical measurements were performed as follows: samples were placed in a cylindrical thermostatic cell consisting of a hollow bronze shell and fluoroplastic cover with mounted aluminum electrode. The shell's bottom utilized as the second electrode. The oil-to-oil thermostat, boiling point of the transformer oil - 220 °C., provided uniform heating of measuring cell. Yokogawa UT750 controller with thermocouple and digital display maintained the required heating rate and accuracy. Laboratory setup for the electrical properties measurements was assembled on NI ELVIS II (National Instruments) hardware-software complex. The temperature dependence of electric conductivity for the samples was obtained in the temperature range of 25 – 165 °C with 20 °C increment. The specimens were kept at the required temperature during 30 minutes before measurements start to avoid adverse inertial effects.

3. Results and discussion

It was supposed, during the origin data study of different asphaltenes and hydrocarbon systems (table 1), that asphaltene fraction is a strong electron donor and acceptor at the same time (IP=4.10 - 6.70 eV, EA=1.80 - 2.50 eV). This means, that the processes in asphaltenes solutions and concentrates, including those related to ARPD (asphalt, resin and paraffin deposits), may be described by the formation of molecular charge-transfer complexes.

Table 1. Electronic structure data of the semiconductors on the base of resins and asphaltenes

Asphaltenes	EIP, eV	EEA, eV	Forbidden gap, eV	Imref, eV
Asphaltenes of Radaevsk oilfield	5.70	1.85	3.85	1.93
Asphaltenes of Surgut oilfield	5.20-5.70	2.10-2.50	3.10-3.20	1.55-1.60
Asphaltenes of straight-run stocks	4.37-5.27	2.44-2.50	1.93-2.77	0.96-1.38
Asphaltenes of tar	4.70-4.90	2.10-2.15	2.60-2.75	1.30-1.38
Asphaltenes of Kushkul oilfield	5.20	1.90	3.30	1.65

The earlier made assumption, that paramagnetic phase of oil asphaltenes belongs to amorphous, compensated, wideband semiconductors is confirmed. AFM study of asphaltenes obtained from crude oil showed the presence of structure fragments ranged from 3 nm to 10 nm, disposed to strong intermolecular interactions. We used different doped compounds for formation of wide gap amorphous semiconductors from a concentrates of asphaltens. Thus, paramagnetic phase of asphaltenes can be used as available semiconductors for nanoelectronics.

The calculations results are confirmed by the EPS spectroscopic data (table 2).

Table 2. Computed and experimental ranges of IP and EA

Data	IP, eV	EA, eV
By UV and VIS spectrums	4.37-5.70	1.85-2.50
Computed molecular fragments (RHF/3-21G** method)	6.36-7.56	0.54-1.65
IP and EA estimation of free-radical fragments	4.92-5.79	1.78-2.45

According to the experimental estimates, IP values are in the range from 4.37 to 5.70 eV, EA - in the range from 1.85 to 2.50 eV. Thus, the best agreement with the experiment is observed for the free radical fragments. Quantum-chemical calculations of molecular and radical fragments with total geometry optimization by molecular mechanics and methods 6-31G** display that the structure of naphtheno-aromatic fragments of asphaltenes is nonplanar and has a "bowl" form (figure 1). The computing of condensed model asphaltene structures shows, that in extreme case of total aromatization of all rings condensed asphaltene fragment have the "plate" structure (figure 1).

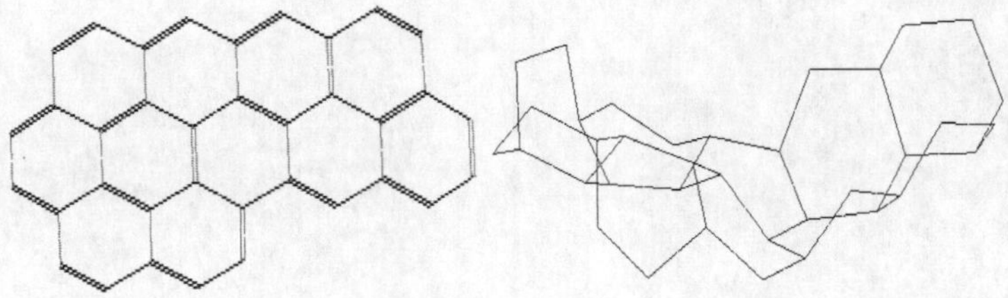

Fig. 1. Condensed models of the asphaltene structure fragments in the cases of total aromatization of the rings and double bonds full absence ("bowl")

There is electrons charge carrier in the asphaltenes [4, 5]. The asphaltenes conductivity were determine of the jump mechanism, as another polyciclic hydrocarbons [4, 5]. The asphaltenes are bulk porous materials, therefore their conductivity depend on external pressure. The asphaltene conductivity less then 10^{-12} S by atmospheric pressure and temperature 298 K. The conductivity increase at rank of the pressure 0 - 10 MPa from 10^{-12} S to 2 10^{-11} S. The conductivity has increased from 10^{-12} S to $2{\cdot}10^{-9}$ S by the temperature variance in the interval from 293 to 453 K. the `Maximum of conductivity was registered for samples oil asphaltenes. For all asphaltenes were observed linear volt-ampere characteristic into interval 0-500V. The conductivity of a DPS substantial increase in comparison with pure asphaltenes. With the increasing reheat temperature from 293 to 453 K, the conductivity values increase from 10^{-12} S to 10^{-5} S for DPS.

This unexpectedly a largeness conductivity value can be explain of the free radicals appearance, that are particulars with an unpaired electron.

Our studies assume that the asphaltenes in the hydrocarbons phase decompose into free radicals, according to the equations [6]:

$$M_n = R^\bullet + R^\bullet_{n-1} \qquad (3)$$

where M_n - the molecular asphaltenes, $R^\bullet$ and $R^\bullet_{n-1}$ - the free radical asphaltenes.

The equilibrium (3) are shifted in right with the temperature increase. The existence of a asphaltenes free radical, as known, is confirmed by ESR data [5, 6]. The concentration of the paramagnetic centers is increased with the grown of temperature from 10^{17} to 10^{19} spin/g. The unpaired electron of free radicals are dope, and so are generated an excess of holes and semiconductors p-type. Moreover, free radicals can miniaturize of band gap , as donors of electron. As a result increases the number of charge carriers greatly within the asphaltenes compounds. Thus, the stable radicals occurrence; provides a phase transition type dielectric-semiconductor.

In an experiments was detected a negative temperature hysteresis of the conductivity. The divergence of hysteresis branches was no more than 30%. The asphaltenes permittivity compiled 5-6 units.

Thus the numerous experiments defined of conductivity testify about phase changes of type dielectric – semiconductor in DPS for areas of 400-500 K. The suggested that of conductivity dynamics is based on the processes of paramagnetic particulas doping of forbidden gap. Experimental results confirm the applicability of this hypothesis for asphaltenes, resins and high-molecular fractions with high concentration of the paramagnetic phase.

Studies of the supramolecular structure of asphaltenes by AFM [3] found that molecular structure of objects with a resolution of about 1 micron is a quasi-gel-like structure consisting of micelle-associated asphaltene molecules. The average concentration of particles in the dispersed phase is 144 particles per 1 μm^2. A more detailed analysis at a resolution of 500 nm (figure 2) shows that the distance between the micelles is about 20-50 nm. The thickness of the particle diameter of 100 nm is less than 3 nm, indicating that stacking structures.

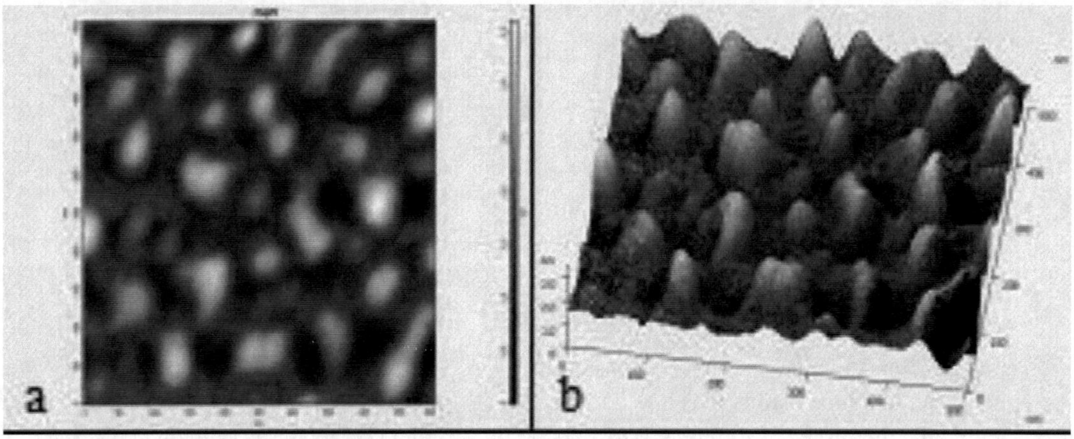

Fig. 2. Results of the supramolecular structure of asphaltenes AFM in detail: a-2D-format b-3D-format

The main of the reaseach conclusion is paramagnetic phase of asphaltenes is organic amorphous broadband semiconductor. Besides this substance can be considere in general as an organic spin glasses.

References

[1] Dolomatov M Yu, Dezortsev S V, Shutkova S A 2012 *J. of Mat. Sci. and Eng. B* **151-157** 2
[2] Dolomatov M Yu, Mukaeva G R, Shulyakovskaya D O 2013 *J. of Mat. Sci. and Eng. B* **183-199** 3
[3] Dolomatov M Yu, Dezortsev S V, Bakhtizin R Z, Shutkova S A, Shulyakovskaya D O, Paymurzina N Kh 2012 *6 th International Meeting on Molecular Electronics* (France) p 160
[4] Evdokimov I N, Losev A P 2010 *Energy&Fuels* **3959-69** 24
[5] Yen T F, Chilingaryan G V 2000 *Asphaltenes and Asphalts 2* (Elsivier)
[6] Unger F G, Andreeva L N 1995 *Fundamental aspects of oil chemistry. Nature of the resins and asphaltenes* (Novosibirsk: Science)

IOP Conf. Series: Materials Science and Engineering **195** (2017) 012006 doi:10.1088/1757-899X/195/1/012006

Enhanced solar light photodegradation of brilliant black bis-azo dye in aqueous solution by F, Sm^{3+} codoped TiO_2

Sabastian S Mukonza*, Edward N Nxumalo, Bhekie B Mamba, Ajay K Mishra*

Nanotechnology and Water Sustainability (NanoWS) Research Unit, College of Science, Engineering and technology (CSET), University of South Africa (UNISA), Florida 1705 Johannesburg, South Africa

E-mails: mishrak@unisa.ac.za, Tel.: +27 116709702

Abstract. This research focuses on improving the photocatalytic efficiency of TiO_2 during the photo-mineralisation of brilliant black (BN) bis-azo dye pollutant in aqueous solution. This was achieved by improving the visible light activity of TiO_2 photocatalyst semiconductor through co-doping of fluorine (F) and trivalent samarium ions (Sm^{3+}) into a TiO_2 matrix using a modified sol-gel synthesis method. Structural, morphological, and textural properties were evaluated using ultra-violet /visible spectroscopy (UV-visible), Raman spectroscopy, scanning electron microscopy coupled to energy dispersive X-ray spectroscopy (SEM/EDX) and X-ray diffraction spectroscopy (XRD). Photocatalytic and degradation efficiencies were assessed by decolourisation of BN dye in aqueous solution. Complete degradation of BN was attained after an irradiation time of 3 h using F, Sm^{3+}-TiO_2 (0.6% Sm^{3+}) compared to 73.4% achieved using pristine TiO_2. Pseudo first order kinetics rate constants (K_a) were 2.73×10^{-2} and 6.6×10^{-3} min^{-1} for Sm^{3+}-TiO_2 (0.6%Sm^{3+}) and pristine TiO_2, respectively, which translates to a remarkably high enhancement factor of 4. The results obtained established that doping of TiO_2 by F and Sm^{3+} enhances the photocatalytic performance of TiO_2 during solar light radiation which enables the utilisation of freely available and clean solar energy.

1. Introduction

Widespread applications of brilliant black (BN) bis-azo dye in food dyeing, auto-mobile painting and textile dyeing industry implies that BN dye has got a higher likelihood of polluting water sources. The presence of the diazo group in BN prompts intolerance to people with allergies to salicylates. Additionally, it is a histamine liberator with a tendency to amplify the symptoms of asthma, its combination with benzoates has got adversarial effects in children leading to hyperactivity and there is sufficient grounds for suspicion that bacterial presence in intestines can decompose them into much more dangerous metabolites.

Semiconductor photocatalysis is an advanced oxidation process (AOP) for complete mineralisation of organic pollutants with more advantages compared to other conventional water treatment methods. TiO_2 which is the commonly used photocatalyst due to its relative affordability, nontoxicity and photo-chemical stability, has got its downsides. These drawbacks includes inactivity in the visible and near infra-red (NIR) region of the solar spectrum and low quantum efficiency resulting from higher carrier charge recombination. TiO_2 multi-doping combination of nonmetals and metals has provided solutions to deal with TiO_2 drawbacks and has been found to consequently improve the rate of photocatalysed reaction. Fluorine doped TiO_2 matrix improves both ultra violet and visible light driven photocatalysis of organic pollutants.

There have been contradictory explanations as to what brings about the photocatalytic enhancement of F-TiO_2 degradation of organic pollutants. This has resulted in the proposition of different models of F-doping effects. Firstly, substitutional doping with F^- either doubly or singly occupying vacant O^{2-} in the TiO_2 matrix as proved by photoluminescence (PL) spectra in work done by Czoska et al [1]. The PL spectra shows that there are two major different types of O^{2-} vacancies namely F and F^* in F doped TiO_2 which results in two different intraband gap states impurity energy levels below the CB at 0.53 and 0.84 eV respectively as shown in figure below [2].

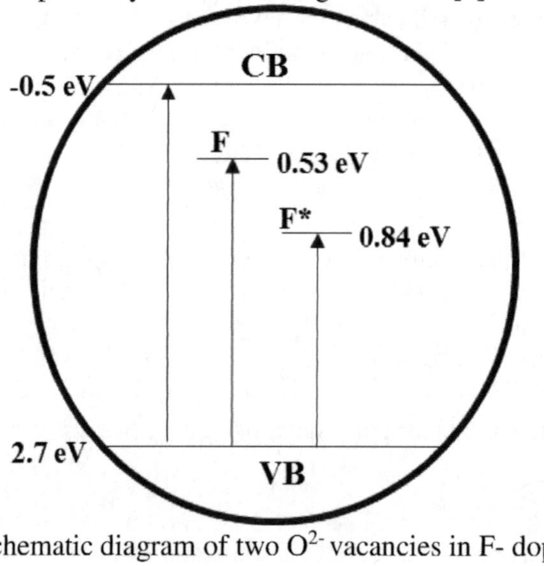

Figure 1: Schematic diagram of two O^{2-} vacancies in F- doped TiO_2

These impurity energy levels results in reduced band gap, an oxygen vacancy induced red shift extending the absorption into the visible region [2] - [3]. Secondly, the presence of singly charged F^{-1} dopant in the doubly charged O^{2-} vacant creates a charge imbalance that is compensated by the formation of Ti^{3+} which could be responsible for the UV activity of F-doped TiO_2. The generated Ti^{3+} reduced centers also results in improved quantum efficiency by acting as electron capture centers for reduced carrier-charge recombination [4]. Thirdly, surface doping by formation of coordinate compounds of Ti^{4+} central metal ion bound by F^- ligands on TiO_2 surface displacing OH^- changing the surface mechanism. This process favours the formation of OH· radicals which in turn increase the kinetic degradation of free radical attacks of organic pollutants [1]. F-doping of TiO_2 alters the surface morphology by enhanced exposition of the {001} facets which are highly photocatalytic active in the UV electromagnetic spectrum [5].

F and metal ion codoped TiO_2 as reported by Lin et al results in improved photocatalytic activity due to the combined effects of F and metal ion on the properties of TiO_2. They found out that the synergistic effects of co-doping F and Ag resulted in extended absorption in the visible region, which could be interpreted as band gap reduction although they did not report calculated band gaps [4]. The presence of Ag also acted as an electron capture center for reduced carrier-charge recombination resulting in improved quantum efficiency.

In metal doping there are mainly four various types of metal dopants that are used namely alkaline [6], transition [7], rare earth [3] and noble [4] metals. Metal dopants function as electron sinks, promotes interfacial charge transfer mechanisms and inhibits recombination of photo-generated carrier charges. The Fermi level for these various metals are lower than the conduction band of TiO_2 therefore photoexcited electrons can be transferred from the CB to the surface of the metal dopant. This on its own is a way of preventing charge carrier recombination [7].

This research focuses on the use of rare earth (RE) metal dopant, Sm^{3+} as a metal dopant in the TiO_2 matrix. The use of RE has generated a lot of research interests due to the unique nature and properties of their 4f electronic orbitals. Due to the higher energy levels possessed by 4f orbitals they can easily duplicate different electronic configurations which are of interesting fluorescence and phosphorescence properties emanating from intra 4f-4f electronic transitions which alters the optical properties of TiO_2 in various ways [8]. This includes among others effective band gap reduction [9], increased absorption intensities in the visible and near infrared (NIR) [9] and conversion phosphors, upper conversion phosphors (UCPs) and lower conversion phosphors (LCPs) [8]. However, the 4f orbitals brings in an added advantage of improved adsorption due to the formation of complex compounds through dative bonding between vacant 4f sub-orbitals and Lewis acids such as functional groups of organic pollutants [10] - [6]. This study also specifically examines the synergistic effects of F and Sm^{3+} codoped TiO_2 for photocatalytic degradation of Brilliant Black (BN) bis-azo dye.

2. Methods and Procedures

2.1 Materials

Titanium isopropoxide (TTIP) 97%, 2-propanol anhydrous 99.5%, formic acid 90%, samarium nitrate hexahydrate ($Sm(NO_3)_3.6H_2O$) 99.9%, ammonium fluoride (NH_4F) 98%, brilliant black 40% and concentrated HCl 32%. All reagents were supplied by Sigma Aldrich (Germany) and were analytical reagent grades which were used without further treatment. In all cases where water was used deionised water (DI) purified in the laboratory using the Milli Quant water purification system was used.

IOP Conf. Series: Materials Science and Engineering **195** (2017) 012006 doi:10.1088/1757-899X/195/1/012006

2.2 Synthesis

2.2.1 Synthesis of fluorine, Sm^{3+} codoped TiO_2 (0.6% Sm^{3+})

F, Sm^{3+} co-doped TiO_2 was prepared by a simple sol-gel method. TTIP (10 mL) was dissolved in 2-propanol (50 mL) and vigorously stirred for 10 min to make **solution A**. A mass of 46.4 mg of samarium nitrate hexahydrate used as Sm^{3+} precursor was weighed and dissolved in a mixture of 2-propanol (10 mL) and DI water (5 mL) (in a ratio of 1:2) by continuously stirring for 30 min to make up **solution B**. The mass weighed gave a 0.6% Sm^{3+} dopant loading w.r.t TiO_2. NH_4F (510.2 mg), F precursor was weighed to give 10% dopant loading of F dopant. NH_4F was dissolved in a mixture of 2-propanol (40 mL), formic acid (10 mL), DI water (3 mL) and HCl (3 mL) and continuously stirred for 30 min to make up **solution C**. Solution C was added dropwise to **solution A** while simultaneously stirring the resultant solution. This was followed by the addition of solution B to the mixture of solutions A and C and vigorously stirred for 1 h. The resultant solution was aged for a period of 24 h and the resultant gel was oven dried over night at 80ºC. The drying oven model DHG 9053A operating at 220 V, 50 Hz and 800 W was used. The pulverized sample was subsequently calcined at 450ºC in a muffle furnace in open air. Muffle furnace model SNOL 8, 2/1100-1 operating between 220-230 V at ~50 Hz and 1,8 KW was used for calcination.

2.2.2 Synthesis of Sm^{3+} doped TiO_2

Sm^{3+} doped TiO_2 was prepared by the same method as explained above but with **solution C** containing formic acid (10 mL) only. **Solution B** comprising of various appropriate amounts of samarium nitrate hexahydrate ranging from 0.2 to 1% dopant loading of each Sm^{3+} while the composition of **solution A** did not change.

2.2.3 Synthesis of pristine TiO_2

Pristine TiO_2 was synthesised using the same sol gel method mentioned in Section 2.2.1. However, in this case TTIP was dissolved in 2-propanol solvent and added dropwise to formic acid (10 mL).

2.3 Characterisation

Optical properties were measured using the PerkinElmer Lambda 1050 UV-visible-NIR. The 150 mm sphere reflectance method was used to scan the samples. D2 and Tungsten lamps were used and the operating parameters slit width of 4 mm and a photomultiplier response of 0.2 seconds. Powdered samples were scanned from 200 – 800 nm range with a data interval of 2 nm.

Raman spectra were collected from a Raman II FT-Raman module spectrometer coupled to a VERTEX 70 FT-IR spectrophotometer, Germany. Resolution of 4 mc^{-1}, sample scan: 64, laser source setting, and scanning range stokes shift 20-700 cm^{-1}.

Crystallinity properties were measured using XRD type Rigaku Smartlab, Japan X-ray diffractometer and the operating parameters were 45 kV and 200 mA. The analysis range was 20 - 80° (2Θ), Cu-Kα radiation wavelength (λ = 1.54 Å) and powdered samples were analysed. Data analysis was carried out using Smartlab Rigaku instrument control software PDXL. Crystallite particle size (D) and lattice strain (η) were calculated using the Scherrer equation, full width half maximum (FWHM) of 101 anatase phase peak.

Surface morphology and elemental composition analysis of the synthesised photocatalyst were performed using the scanning electron microscopy (SEM) with an attachment of energy dispersive X-ray spectrometer called JOEL IT 300 & Oxford SEM Integrated System.

Evaluation of photocatalytic activity

The photocatalytic activity of Sm^{3+} doped TiO_2 (0 – 1%), F doped TiO_2 and F, Sm^{3+} codoped TiO_2 was evaluated by measuring the photocatalytic degradation of brilliant black (BN) bis-azo dye under solar light irradiation. For every sample 0.1 g of photocatalyst in 100 mL of 50 ppm BN bisazo dye was mixed and stirred in darkness for 1 h to attain absorption equilibria before solar light irradiation. Solar simulator (HAL – 320) supplied by ASAHI SPECTRAL, Japan was used to generate the required spectral irradiance for photocatalysis. The 300 W compact Xenon lamp was the light source, and a light intensity of 60 was used. Solar simulator radiation spectrum output wavelength range was set from 350 – 1100 nm and an air mass filter 1.5 G filter was used to filter the unwanted wavelength. Illuminance distance, which is distance from collimator lens to sample surface was approximately 370 mm creating an effective radiated area of 50 mm^2 and an irradiance of 100 $mWcm^{-2}$ (1 SUN) . The simulator was warmed up 30 min prior to application and was used as solar light source for irradiation. Sampling frequency was after every 30 min interval which also included initial sample after attaining absorption equilibria, before solar light irradiation. Samples of 4 mL aliquots were drawn using a 10 mL syringe were filtered through the GH Polypro membrane (GHP) syringe disc filter fitted at the discharge of the syringe. Filtered samples were analysed for absorbance using the UV-visible spectrophotometer at wavelength 574.9 nm. Calibration range of 10 to 50 ppm standards was performed which gave R^2 of 0.9999997 and regression line:

$$y = 0.044994x - 0.003554 \qquad \text{Equation 1}$$

where y is absorbance and x is the BN concentration in ppm.

This equation was used to determine the BN concentration for every sample and the BN concentrations at different sampling times were used to calculate the percentage degradation of BN using the following formula:

$$\%\ \text{Degradation of BN} = \frac{(C_o - C_t)}{C_o} \times 100 \qquad \text{Equation 2}$$

where C_o and C_t are initial concentration and concentration at time t respectively.

The degradation progress was monitored using total organic carbon (TOC) torch analysis wastewater method. Total organic carbon (TOC) analysis which basically measures the amount of organic carbon in the samples was done in a Teledyne Tekmar TOC torch (USA). Standard solutions of 1 ppm, 5 ppm, 10 ppm, 20 ppm and 30 ppm carbon were prepared using potassium hydrogen phthalate (KHP) and DI water. Calibration standards were measured prior to sample analysis for instrument calibration.

3. Results and Discussion

UV-visible reflectance spectra showed the optical activity of various synthesised samples of pristine TiO_2 and various doped samples. For pristine and Sm^{3+} doped TiO_2 there was a slight shift in the band gap, as Sm^{3+} dopant loading increased the band gap was progressively narrowed but Sm^{3+} doped TiO_2 showed varying and increased absorption intensity spanning from 400 to 800 nm. This increased absorption intensity extends to the NIR region of the electromagnetic spectrum. This phenomenon also persists in the UV region as shown in Figure 2. Fluorine doped TiO_2 and F, Sm^{3+} codoped TiO_2 showed significant shifts of the band gap. Converting diffuse reflectance data to Kubelka-Munk function using the following equation below:

$$F(R) = \frac{(1 - R)^2}{2R} \qquad \text{Equation 3}$$

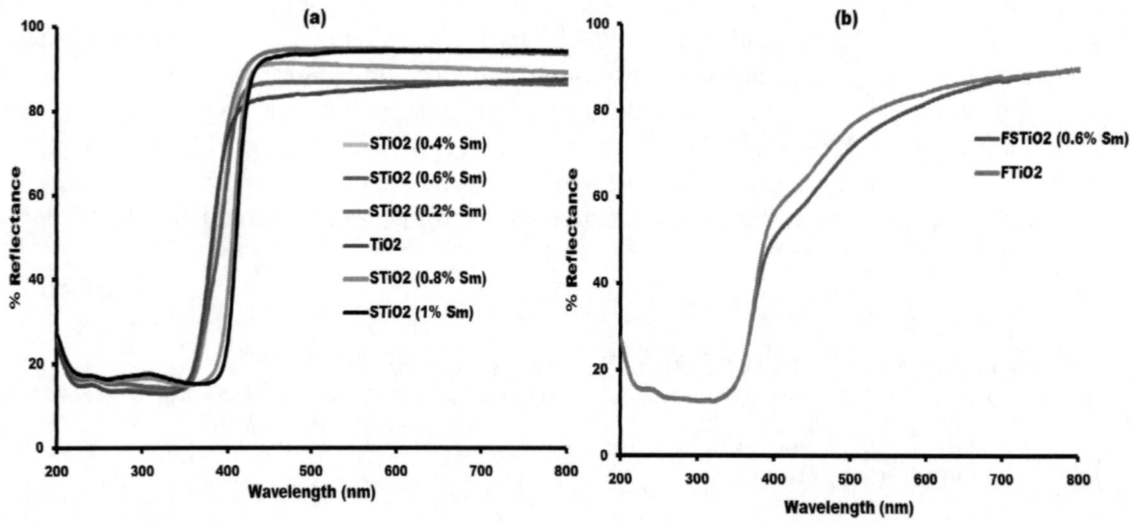

Figure 2: Diffuse reflectance UV-visible spectra for doped TiO_2 (a) different Sm^{3+} concentrations and (b) F-doped TiO_2 and F, Sm^{3+} codoped TiO_2 (0.6% Sm^{3+})

The Kubelka-Munk function, F(R) is an absorption coefficient showing the magnitude of absorption for a sample. F(R) is plotted against wavelength to show different absorption behaviour of different photocatalysts as shown in Figure 3 which confirms reflectance data shown by UV-visible spectra. Further treatment of the reflectance data to Tauc plots gave a good estimate for indirect band gap energies for pristine TiO_2 and Sm^{3+} doped TiO_2 showing the indirect band gaps

IOP Conf. Series: Materials Science and Engineering **195** (2017) 012006 doi:10.1088/1757-899X/195/1/012006

in Tables 1 and 2 below. Tauc plots below (Figure 4) plots point of $[F(R) \times h\nu]^{1/n}$ against hv where hv is photon energy and n is the numerical value corresponding to type of transition. The values of n can be ½, 2, 3/2, 3 corresponding to allowed direct, allowed indirect, forbidden direct or forbidden indirect transition. In the case of TiO_2 n=2 to estimate the indirect band gap.

This is largely attributed to phosphorescence and fluorescence optical properties of lanthanides trivalent ions. The unique electronic configuration of Sm^{3+} consisting of partially filled 4f electronic orbitals responsible for distinctive optical transitions behaviour of lanthanides. The increased absorption intensity in UV, visible to NIR by the Sm^{3+} doped TiO_2 is due to weak, narrow sharp optical absorption by Sm^{3+}. This narrow sharp optical absorption is attributed to intra 4f-4f electronic transitions in Sm^{3+} [11].

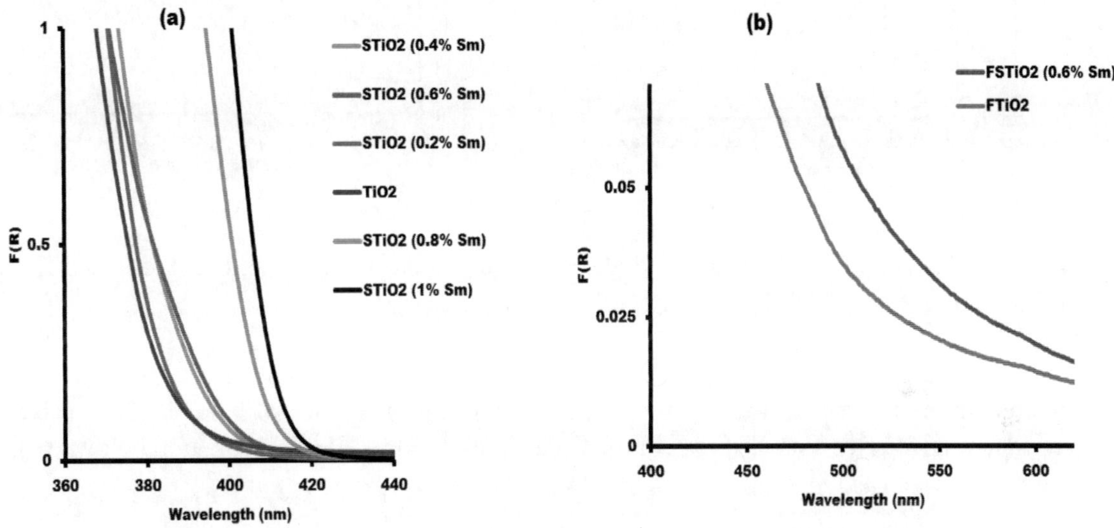

Figure 3: Kubelka-Munk function plots for doped TiO_2 (a) different Sm^{3+} concentrations and (b) F-doped TiO_2 and F, Sm^{3+} codoped TiO_2 (0.6% Sm^{3+})

While pristine and Sm^{3+} doped TiO_2 showed a slight band gap narrowing, the doping of TiO_2 with nonmetal dopants F has shown significant band gap narrowing as shown in Figure 4 below. Doping TiO_2 with F and Sm^{3+} narrowed the band gap of TiO_2 and consequently increased the wavelength absorption edge of doped TiO_2 into the visible region of the electromagnetic spectrum as shown in Table 1 and 2 below. This is because the introduction of F and Sm^{3+} introduced impurities which in turn introduced intra-band gap energy states in between the TiO_2 VB and CB. F^- anionic radius is much less than that of the O^{2-} (1.33 Å < 1.4 Å) [4], so there is a high probability that F^- occupies a substitutional doping site replacing O^{2-} resulting in a 2p energy sub-state above the VB, while Sm^{3+} is much larger than O^{2-} (0.96 Å >> 0.68 Å) [12], and is more likely to occupy the interstitial doping site or substitute Ti^{4+} resulting in a 3p energy sub-states slightly below the CB [13]. Below Figure 10 shows the possible band gap modification of TiO_2 by doping with F, Sm^{3+} codoped TiO_2.

IOP Conf. Series: Materials Science and Engineering **195** (2017) 012006 doi:10.1088/1757-899X/195/1/012006

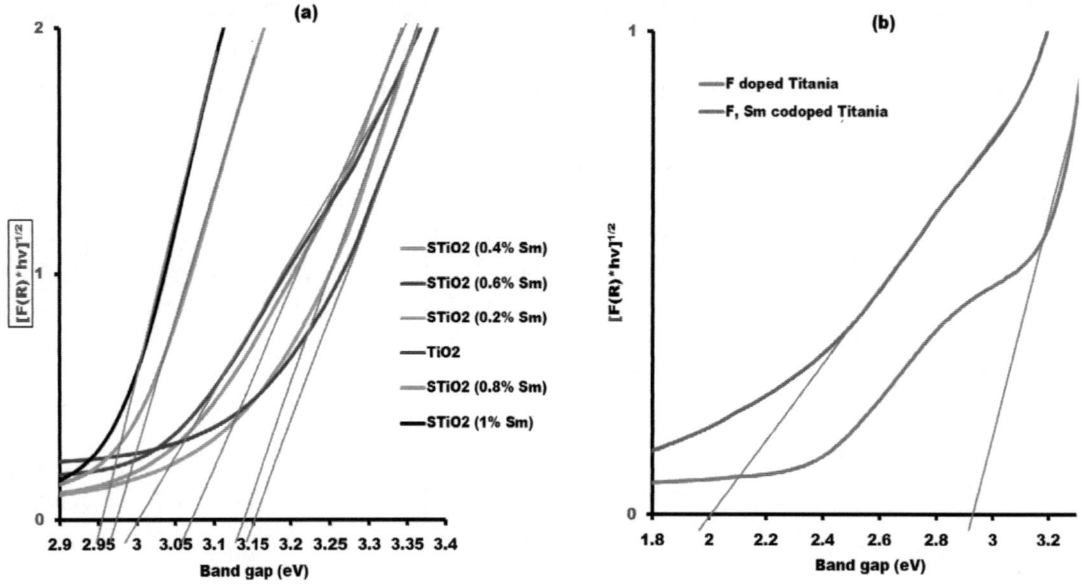

Figure 4: Tauc plots for doped TiO_2 (a) different Sm^{3+} concentrations and (b) F doped TiO_2 and F, Sm^{3+} codoped TiO_2 (0.6% Sm^{3+})

As illustrated below in Table 1 and 2 below, the estimated band gap energies and calculated wavelength for pristine and the different doped TiO_2 samples are shown. The calculated wavelength were derived from the Planck-Einstein relationship giving the Planck's equation below [14]:

$$Eg = h\nu = \frac{hc}{\lambda} = \frac{1240}{\lambda} \qquad \text{Equation 4}$$

Where h is Planck's constant, v is frequency, c is speed of light and λ is wavelength

Table 1: Indirect band gap energy estimates and calculated absorption wavelength for different Sm^{3+} dopant concentrations

Sample	Indirect band gap (eV)	Estimated λ (nm)
Pure TiO_2	3.15	393.7
Sm^{3+}-TiO_2 (0.2% Sm^{3+})	3.14	394.9
Sm^{3+}-TiO_2 (0.4% Sm^{3+})	3.07	403.9
Sm^{3+}-TiO_2 (0.6% Sm^{3+})	3.00	413.3
Sm^{3+}-TiO_2 (0.8% Sm^{3+})	2.97	417.5
Sm^{3+}-TiO_2 (1% Sm^{3+})	2.95	420.3

IOP Conf. Series: Materials Science and Engineering **195** (2017) 012006 doi:10.1088/1757-899X/195/1/012006

Table 2: Indirect band gap energies and calculated absorption wavelength for F-doped TiO_2 and F and Sm^{3+} codoped of TiO_2 on band gap and estimated wavelength

Sample	Indirect band gap (eV)	Estimated λ (nm)
F-TiO_2	2.95	420.3
F, Sm^{3+}-TiO_2 (0.6% Sm^{3+})	2.00	620.0

Figure 5 below shows Raman spectra of pure TiO_2 and Sm^{3+} doped TiO_2 with well-defined high intensity sharp E_g peaks at 146 cm^{-1} and 646 cm^{-1}, these peaks confirms the presence of crystalline lattices of TiO_2. Peaks B_{1g} (397 cm^{-1}) and A_{1g}+B_{1g} (516 cm^{-1}) are also well defined but with low Raman intensity and broader peaks and an offshoot E_g peak at 197 cm^{-1} is observed [15]. All these peaks confirms the presence of a predominantly anatase TiO_2 phase and these findings are consistent with the XRD spectra data. The presence of the various Sm^{3+} dopants in TiO_2 can be explained by the shifting of peak wavenumber coordinate points along the x-axis [16].

Widespread literature survey reported the main E_g peak to be at 146 cm^{-1} for pure TiO_2 this is exactly what is experimentally obtained in this research while its corresponding Sm^{3+} doped TiO_2 sample had this same peak translated to 149 cm^{-1}, this major shift may represent the presence of Sm^{3+} dopants. Raman active properties of Sm^{3+}-TiO_2 compounds are mainly due to O-Sm-O bending appearing in the region of 304 cm^{-1} and 308 cm^{-1} (E_g and F_{2g} modes) and Sm-O stretching appearing in the region of 515 cm^{-1} and 519 cm^{-1} (A_{1g} mode). These various Raman active modes peaks of Sm^{3+}-TiO_2 compounds are not visible due to low intensity and overlapping of major TiO_2 peaks appearing in the same region. Low intensity is due to low Sm^{3+} dopant concentration [16].

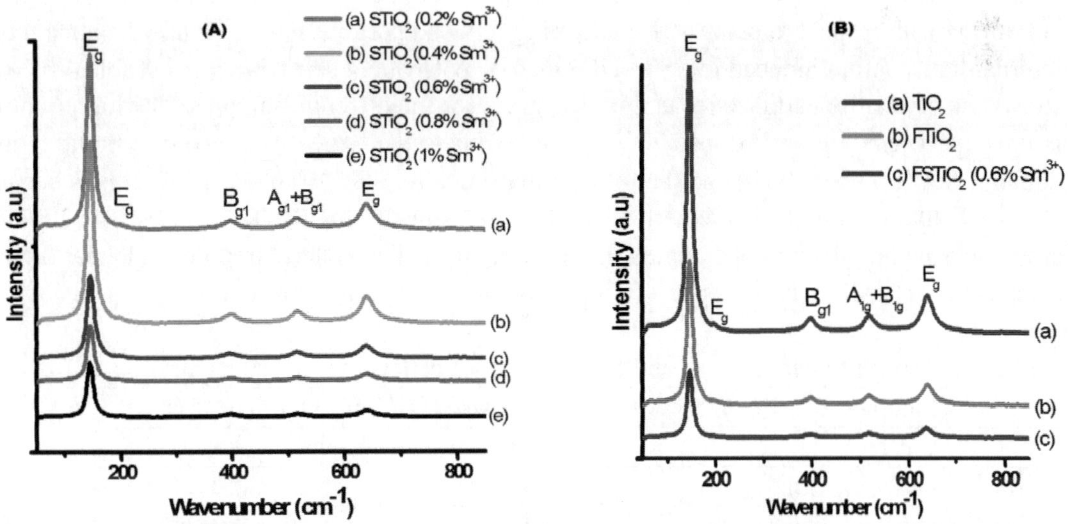

Figure 5: Raman spectra of (A) TiO_2 doped with different Sm^{3+} concentrations and (B) pure TiO_2, F-TiO_2 and F, Sm^{3+} codoped TiO_2 (0.6% Sm^{3+})

The effect of Sm^{3+} concentration is shown in figure (A), intensity of all the anatase peaks was significantly reduced as the Sm^{3+} dopant concentration levels are increased. This confirms reduced crystallinity as the amount of dopant is also increased and this finding is corroborated by XRD crystallite particle size results.

Both pristine and different concentrations Sm^{3+} doped TiO_2 were calcined at 450ºC. At 450ºC calcination temperature, sharp well defined, intense peaks of anatase polycrystalline structure at different 2Θ angle positions were observed with the major peak being at 25.3° position. The observations were made for both pristine and Sm^{3+} doped TiO_2. These peaks showed a higher degree of crystallinity. Other anatase diffraction peaks existing at 2Θ position 25.3°, 37.6°, 48.2°, 53.9°, 54.8°, 62.7° and 75.2° shown in figure 6 below. These diffraction peaks corresponding to (101), (004), (200), (105), (211), (204) and (215) crystallographic planes respectively [13]. Three low intensity peaks of rutile TiO_2 are seen at 27.5°, 41.3° and 56.7° corresponding to crystallographic plane (110), (111) and (220) [17]. The rutile peaks disappeared as the concentration of Sm^{3+} dopant was increased. This implies that Sm^{3+} doping inhibited phase transformation from anatase to rutile. Anatase remained the dominant phase after doping as shown by the XRD peaks. Crystallite particle size (D) and lattice strain (η) were calculated using the formula below:

$$D = \frac{k\lambda}{\beta Cos\theta} \qquad \text{Equation 5}$$

where K is a constant 0.94, λ is the characteristic X-ray wavelength of Cu 1.54 Å, β is full width half maximum (FWHM) for plane 101 and Θ is the Bragg angle in radians. Lattice strain (η) is obtained from online XRD calculator. Introduction of dopant Sm^{3+} broadened the width measured at FWHM with respect to plane 101 confirming reduced particle size. This may be attributed to internal lattice strain induced into the TiO_2 matrix by trivalent samarium [16]. Table 4 below shows the crystallite particle size and lattice strain for the different samples. The higher lattice strain experienced by Sm^{3+} doped TiO_2 might be due to the large difference in cationic radii of the dopant (Sm^{3+}: 0.96 Å) [18] and that of host matrix cation (Ti^{4+}=0.68 Å) [16]. This is because when substitutional doping is experienced, the Sm^{3+} substituting the Ti^{4+} causes both strain and stress to bond length and bond angles (lattice straining). The general trend reveals that there is a decrease in crystallite particle size of TiO_2 with increasing Sm^{3+} concentration.

Table 3: Crystallite particle size and lattice strain of different Sm^{3+} concentration

Sample	Crystallite particle size (D)/ nm	Lattice strain (η)
Pure TiO_2	15.14	0.0119
Sm^{3+}-TiO_2 (0.2% Sm^{3+})	14.9	0.0111
Sm^{3+}-TiO_2 (0.4% Sm^{3+})	14.4	0.0115
Sm^{3+}-TiO_2 (0.6% Sm^{3+})	14.18	0.0117
Sm^{3+}-TiO_2 (0.8% Sm^{3+})	13.29	0.0124
Sm^{3+}-TiO_2 (1% Sm^{3+})	12.84	0.0140

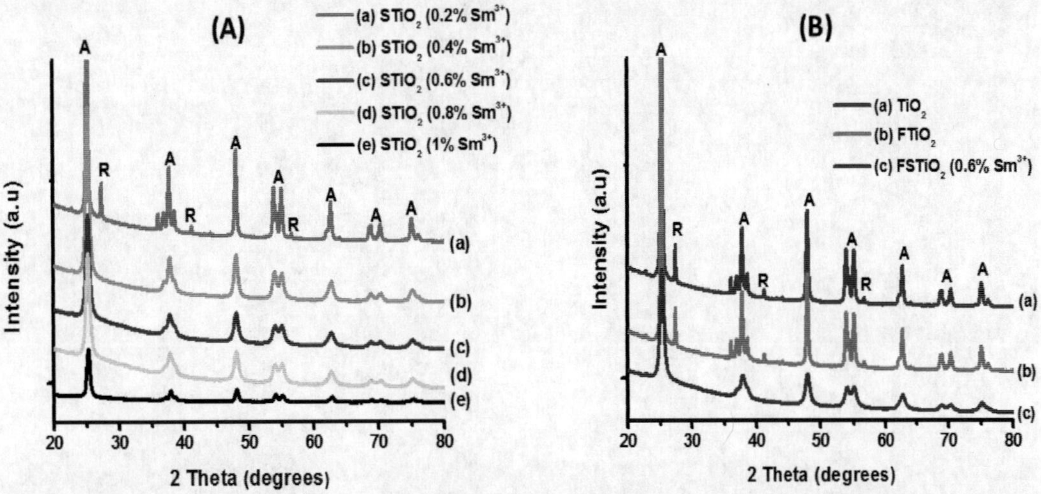

Figure 6: XRD spectra of (A) TiO_2 doped with different Sm^{3+} concentrations and (B) pure TiO_2, F-TiO_2 and F, Sm^{3+} codoped TiO_2 (0.6% Sm^{3+})

SEM micrographs digital pictorial images of surface morphology and surface elemental composition of F, Sm^{3+} codoped TiO_2 shown in Figure 7 below (a) which is high magnification at x2000 and (b) at lower magnification at x450. At lower magnification the digital photo shows an aerial view of small spherical particles with an off whitish colour confirming TiO_2 as the major component of the sample and this is corroborated by the EDX spectra. On zooming in closely the picture revealed further detailed information as shown in picture (a), showing homogeneously dispersed regular spherical and near spherical particles. Figure 7(b) show neither agglomeration nor aggregation of titanium dioxide particles. Figure (c) is EDX spectra showing Ti and O as the major components of the sample. Minute quantities of F and Sm confirms their presents as co-dopants. Sm and Ti have three or more peaks each appearing at different binding energies confirming that both Sm and Ti exists in different oxidation states in the catalyst. However, Si peak shows the presence of minute Si as a contaminant probably from borosilicate glassware used during synthesis. The higher counts per second of the carbon peak is due to the carbon tape which was used as a sample holder.

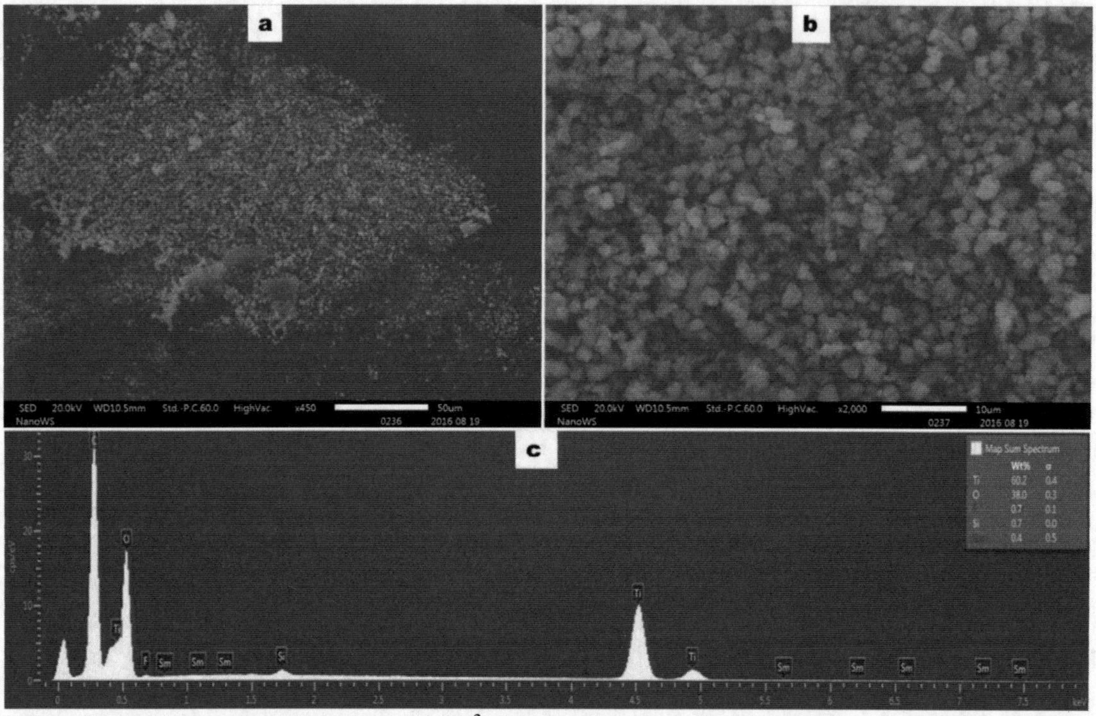

Figure 7: SEM micrographs of (a) F, Sm^{3+} codoped TiO_2 at low magnification (x450), (b) F, Sm^{3+} codoped TiO_2 at higher magnification (x2000) and (c) EDX spectra for F, Sm^{3+} codoped TiO_2

3.1 Dye degradation

Figure 8 shows the ratio of concentration of the dye present at time (t) to initial dye concentration of BN bis-azo dye as a function of irradiation time. A 100% degradation of BN was realized after 3h irradiation for the F/Sm^{3+}-TiO_2 (0.6% Sm^{3+}) sample. The plots shows different rates of degradation for various TiO_2 samples. Pristine TiO_2 has the least photocatalytic activity, however, the percentage degradation of BN dye by pristine TiO_2 under solar light irradiation is significantly higher (73.8%) which might be due to either dye sensitisation or BN bis-azo dye being easier to photodegrade. F- doped TiO_2 marginally increased the photocatalytic degradation of BN to 78.4%.

After 180 min of solar light illumination, different Sm^{3+} doped TiO_2 samples showed remarkable enhancement of degradation relative to pristine TiO_2 which may be attributed to the improved quantum efficiency with Sm effectively acting as an electron capturing centres for photoinduced electrons. This avoided recombination with positive holes and improved the photocatalytic activity. Schottky barriers formed at the semiconductor-Sm^{3+} interface also acted as electron sinks reducing recombination rate. Increasing Sm^{3+} concentration up to 0.6%, resulted in increased photocatalytic efficiency of TiO_2 but a further increase in Sm^{3+} concentration resulted in diminishing activity. The diminishing activity at higher concentrations of Sm^{3+} is due to Sm^{3+} acting as recombination centers further facilitating recombination by short circuiting the electrical circuit which is detrimental to photocatalytic improvement. Increased Sm^{3+} concentration on the TiO_2 surface leads to shading of photo-sensitive reaction sites of TiO_2 by the metal cations. Optimum metal dopant loading for different Sm^{3+} was found to be 0.6%.

The photocatalytic activity of BN bis-azo dye can also be described by the Langmuir-Hinshelwood model, where the photocatalytic activity of the catalysts can be quantitatively evaluated by Equation 6 [19].

$$Ln\left(\frac{C_o}{C_t}\right) = kt \qquad \text{Equation 6}$$

where k is the apparent reaction rate constant and C_0 and C_t represent the concentrations at the initial time and time t, respectively.

The kinetics of BN degradation is presented in Figure 9 and Table 4. Plotting the logarithm of the normalized dye concentration against irradiation time for 180 min gave a good approximation of the rate constants. Nearly perfect linear relationships with high value of R^2 were observed, indicating that all reactions followed pseudo first order kinetics. The kinetics data shows that F, Sm^{3+} codoped TiO_2 (0.6% Sm^{3+}) has the fastest rate of reaction compared to all the other photocatalysts. This is supported by the trends observed in photodegradation rate and percentage TOC removal.

Comparing the photocatalytic activity of the synthesised F, Sm^{3+} codoped TiO_2 with data in literature showed that F, Sm^{3+} codoped TiO_2 performed relatively well. Sun and co-workers reported a k value of 1.6×10^{-3} min^{-1} for N-doped TiO_2 prepared at a pH of 5.5 [20]. While, Kuvarega et al. synthesised N/Pd doped TiO_2 (0.0% Pd), N/Pd-codoped TiO_2 (0.6% Pd) which showed a K_a value of 2.35×10^{-3} min^{-1} with highest initial rate (1.86×10^{-2} min^{-1}) for the degradation of Eosin yellow [13]. The values reported by the two authors showed much lower catalytic activity compared to findings of the present work. F-doped TiO_2 and F, Sm^{3+} codoped TiO_2 (0.6% Sm^{3+}) showed K_a values of 8.6×10^{-3} and 2.73×10^{-2} min^{-1} respectively which were higher than those reported by the two authors.

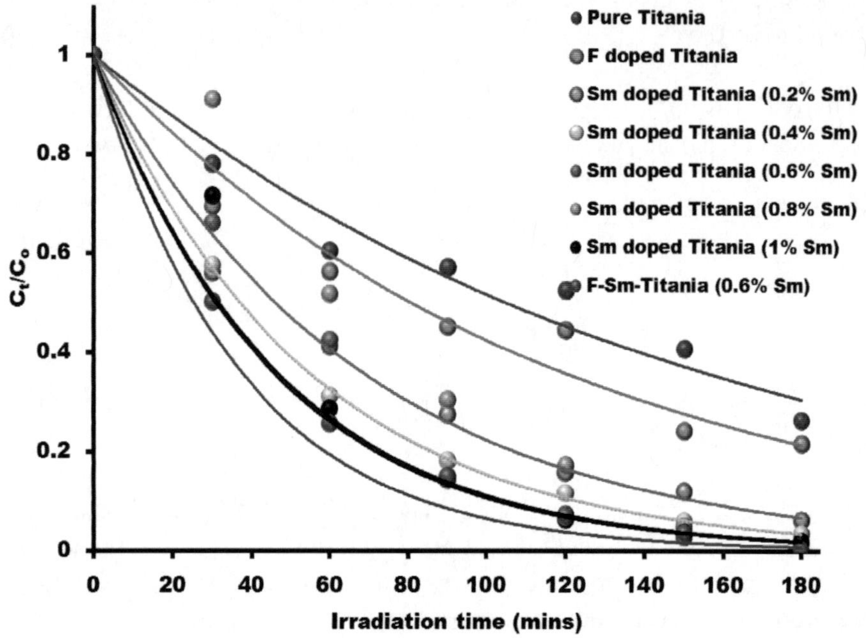

Figure 8: BN bis-azo dye degradation efficiency of various photocatalysts

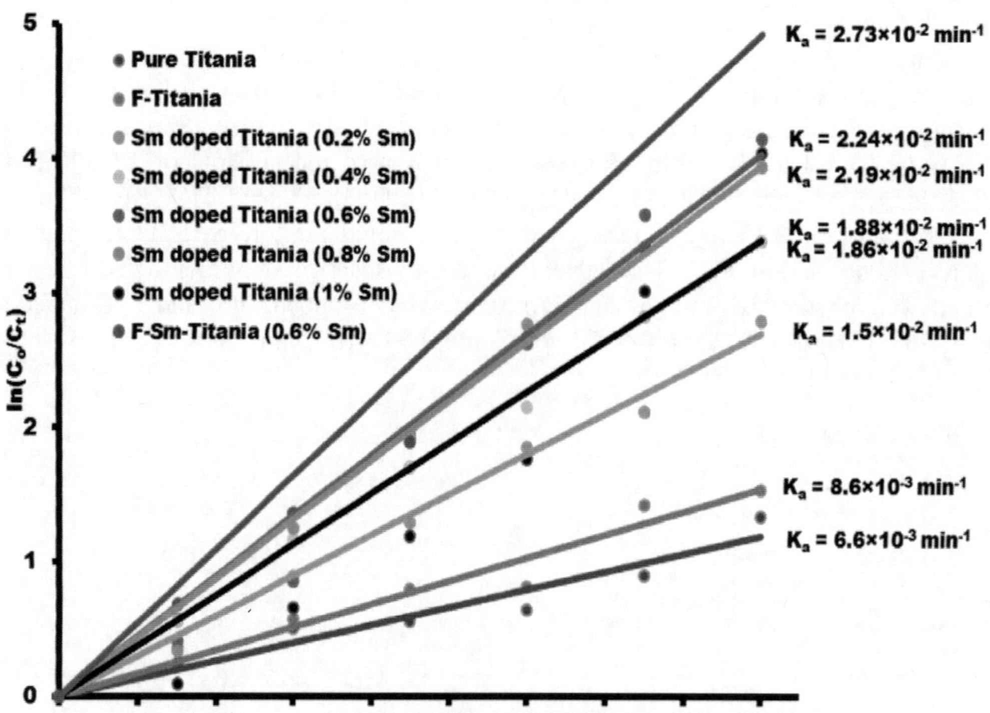

Figure 9: Pseudo first order kinetics for degradation of BN bis-azo dyes for 180 min

Table 4 shows the rate constants for different TiO_2 samples after 3 h of irradiation with solar light. F, Sm^{3+} codoped TiO_2 (0.6% Sm^{3+}) showed the highest degradation rate and the highest K_a value due to the synergistic effects.

Table 4: Degradation rates and pseudo first order rate constants after 3 h solar irradiation

Sample	Degradation after (%)	Rate constant (K_a)/min^{-1}
Pure TiO_2	73.8	6.6×10^{-3}
F-TiO_2	78.4	8.6×10^{-3}
Sm^{3+}-TiO_2 (0.2%)	93.8	1.5×10^{-2}
Sm^{3+}-TiO_2 (0.4%)	96.6	1.86×10^{-2}
Sm^{3+}-TiO_2 (0.6%)	98.4	2.24×10^{-2}
Sm^{3+}-TiO_2 (0.8%)	98.2	2.19×10^{-2}
Sm^{3+}-TiO_2 (1%)	98	1.88×10^{-2}
F-Sm^{3+}-TiO_2 (0.6%)	100	2.73×10^{-2}

Figure 10 shows the proposed mechanism for the synergistic effects of F and Sm^{3+} on TiO_2. Fluorine and Samarium have synergistic effects responsible for improving the photocatalytic activity of TiO_2 under visible light irradiation. Under visible light irradiation, substitutional F dopant creates two intra band gap states below to the conduction band edges leading to a narrower band gap than pristine TiO_2. The schematic diagram presented in Figure 10 shows the proposed possible mechanism for the synergistic effects of fluorine and samarium on the codoped photocatalysts.

Under light irradiation, electrons (e^-) are photo-excited from the valence band (VB) to the conduction band (CB), creating positive holes (h^+) in the VB. The introduction of a substitutional or interstitial fluorine in the TiO_2 results in the formation of intra band gap states close to the valence band. This enhances the visible light activity of the nanoparticles as electrons can be excited from the fluorine 2p orbitals to the conduction band. In pristine TiO_2, the generated carrier charges recombines and only a smaller fraction of the electrons and holes participate in the photocatalytic reaction, resulting in low photocatalytic activity. In the presence of Sm, the photogenerated electrons are trapped, leading to electron-hole separation. The holes scavenge surface adsorbed water or hydroxyl molecules generating highly reactive hydroxyl radical species. The electrons, on the other hand, scavenged the oxygen molecules to form very reactive superoxide radicals. Both radicals are highly reactive towards the degradation of the azo dye compounds and other organic compounds such as BN dye.

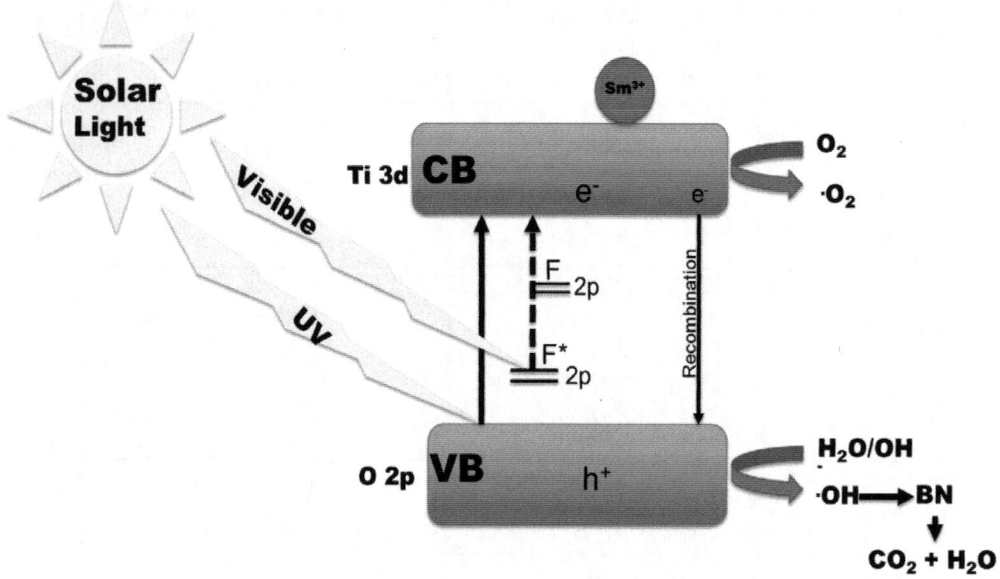

Figure 10: Proposed synergistic effects of F and Sm^{3+} doping on TiO_2

Figure 11 (a) and (b) shows colour disappearance and the disappearance of the absorption peak at 574.9 nm absorption wavelength. This is evidence that that the chromophere responsible for absorption of visible light that is characteristic colour of the dye was being split into smaller and simpler molecules.

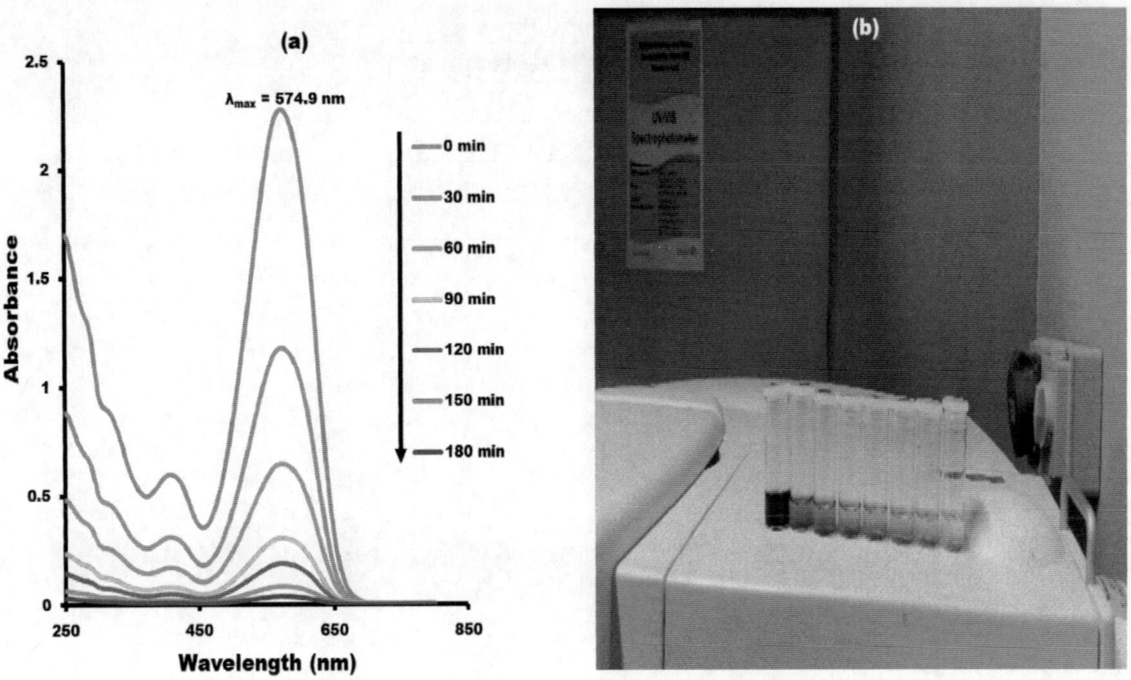

Figure 11: Degradation profile for BN dye using F, Sm^{3+} codoped TiO_2 (0.6% Sm^{3+}) (a) UVvisible absorption spectra and (b) digital photograph showing decolourisation of BN dye.

Monitoring dye degradation profile is important to establish the degree of mineralisation and products of photodegradation. This is because some intermediate products like nitrated poly aromatic compounds are toxic, mutagenic, teratogenic and carcinogenic therefore not desirable [21]. Complete mineralisation of dye molecules to simpler and non-toxic compounds like CO_2, H_2O and inorganic ions is ideally desirable [14]. TOC analysis was used to monitor the organic carbon bound in the BN molecule as an indicator to complete mineralisation of BN dye. Figure 12 and Table 5 below show initial TOC concentration of 15 ppm and removal of 82% corresponding to 2.7 ppm concentration of TOC after 3 h irradiation time. This was significantly a higher fraction of TOC removal but the TOC concentration at 2.7 ppm was well above the maximum permissible limit which is 0.05 ppm. A further increase in degradation time to 4 h achieved complete TOC removal which means that BN dye was completely mineralised to simpler less toxic molecules and ions. The initial total organic concentration was 15 ppm and the final concentration after 4 h was 0 ppm.

IOP Conf. Series: Materials Science and Engineering **195** (2017) 012006 doi:10.1088/1757-899X/195/1/012006

Table 5: TOC removal during 4 h irradiation time

Irradiation time (h)	TOC (ppm)	% TOC removal
0	15.0	-
1	5.6	62.4
2	4.2	71.7
3	2.7	82
4	0	100

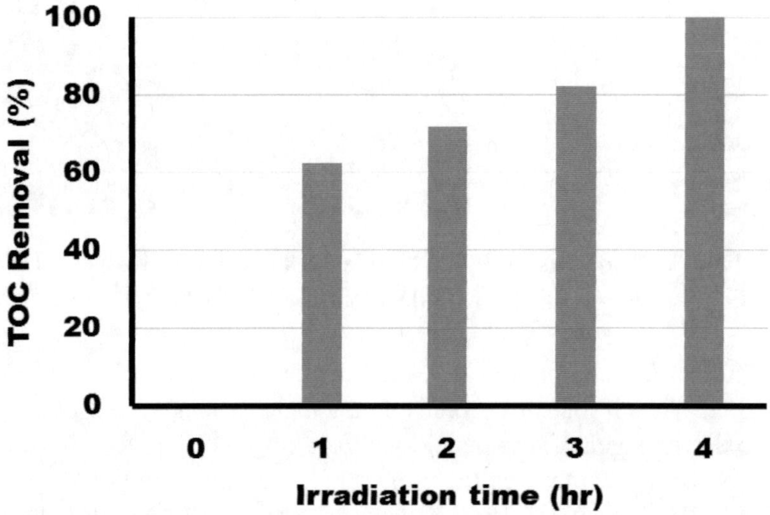

Figure 12: The extent of TOC removal after 4 h

4. CONCLUSION

F, Sm^{3+} co-doped TiO_2 was successfully synthesised by the sol-gel method and was open air calcined at 450 °C. F, Sm^{3+} co-doped TiO_2 showed crystalline particles of predominantly anatase phase which were solar responsive. F and Sm^{3+} codoping significantly reduced TiO_2 band gap and improved the quantum efficiency of the photocatalyst. This significantly improved the photocatalytic effect of F, Sm^{3+} co-doped TiO_2, relative to that of pristine TiO_2 by an enhancement factor of 4. The different photocatalytic reactions followed pseudo first order kinetics with higher R^2. A higher reaction rate resulted in complete degradation of BN dye after 3 h of solar irradiation. TOC confirmed complete dye mineralisation after irradiation of 4 h, an indication that there was low probability of producing toxic metabolites. This is a successful, effective and innovative watertreatment method for the remediation of BN in aqueous water.

REFERENCES

[1] A.M. Czoska, S. Livraghi, M. Chiesa, E. Giamello, S. Agnoli, G. Granozzi, et al., The Nature of Defects in Fluorine-Doped TiO_2, J. Phys. Chem C. 112 (2008) 8951–8956.

[2] H. Sun, S. Wang, H.M. Ang, M.O. Tadé, Q. Li, Halogen element modified titanium dioxide for visible light photocatalysis, Chem. Eng. J. 162 (2010) 437–447.

[3] L. Yang, P. Liu, X. Li, S. Li, The photo-catalytic activities of neodymium and fluorine doped TiO_2 nanoparticles, Ceram. Int. 38 (2012) 4791–4796.

[4] X. Lin, F. Rong, D. Fu, C. Yuan, Enhanced photocatalytic activity of fluorine doped TiO_2 by loaded with Ag for degradation of organic pollutants, Powder Technol. 219 (2012) 173–178.

[5] E.M. Samsudin, S.B.A. Hamid, J.C. Juan, W.J. Basirun, G. Centi, Enhancement of the intrinsic photocatalytic activity of TiO_2 in the degradation of 1,3,5-triazine herbicides by doping with N,F, Chem. Eng. J. 280 (2015) 330–343.

[6] F. Han, V.S.R. Kambala, M. Srinivasan, D. Rajarathnam, R. Naidu, Tailored titanium dioxide photocatalysts for the degradation of organic dyes in wastewater treatment: A review, Appl. Catal. A Gen. 359 (2009) 25–40.

[7] A.T. Kuvarega, R.W.M. Krause, B.B. Mamba, Multiwalled carbon nanotubes decorated with nitrogen , palladium co-doped TiO_2 (MWCNT/N , Pd co-doped TiO_2) for visible light photocatalytic degradation of Eosin Yellow in water, J Nanopart Res. 2 (2012) 776.

[8] J. Reszczyńska, T. Grzyb, J.W. Sobczak, W. Lisowski, M. Gazda, B. Ohtani, et al., Lanthanide co-doped TiO_2: The effect of metal type and amount on surface properties and photocatalytic activity, Appl. Surf. Sci. 307 (2014) 333–345.

[9] Václav Stengl, S. Bakardjieva, N. Murafa, Preparation and photocatalytic activity of rare earth doped TiO_2 nanoparticles Mater. Chem. Phys. 114 (2009) 217–226.

[10] C.H. Liang, F.B. Li, C.S. Liu, J.L. Lü, X.G. Wang, The enhancement of adsorption and photocatalytic activity of rare earth ions doped TiO_2 for the degradation of Orange I, Dye. Pigment. 76 (2008) 477–484.

[11] Y. Tang, W. Di, X. Zhai, R. Yang, W. Qin, NIR-Responsive Photocatalytic Activity and Mechanism of $NaYF_4$, Tm@TiO_2 Core-Shell Nanoparticles, ACS Catal. (2013) 405–412.

[12] L. Lin, W. Lin, J.L. Xie, Y.X. Zhu, B.Y. Zhao, Y.C. Xie, Photocatalytic properties of phosphor-doped titania nanoparticles, Appl Cat B Enviro. 75 (2007) 52–58.

[13] A.T. Kuvarega, R.W.M. Krause, B.B. Mamba, Nitrogen / Palladium-Codoped TiO_2 for Efficient Visible Light Photocatalytic Dye Degradation, J.Phys.Chem.C. (2011) 22110–22120.

[14] G. Mamba, X.Y. Mbianda, A.K. Mishra, Enhanced visible light photocatalytic degradation of eriochrome black T and eosin blue shade in water using tridoped titania decorated on SWCNTs and MWCNTs : Effect of the type of carbon nanotube incorporated, Mater. Chem. Phys. 149–150 (2015) 734–742.

[15] G. Mamba, X Mbianda, A Mishra, Gadolinium nanoparticle-decorated multiwalled carbon nanotube / titania nanocomposites for degradation of methylene blue in water under simulated solar light, Env. Sci Pollut Res. 21 (2014) 5597–5609.

[16] J. Navas, A. Sánchez-coronilla, T. Aguilar, D.M.D.L. Santos, N.C. Hernández, R. Alcántara, et al., Thermo-selective $Tm_xTi_{1-x}O_{2-x/2}$ nanoparticles: from Tm-doped anatase TiO_2 to a rutile/pyrochlore Tm2Ti2O7 mixture. An experimental and theoretical study with a photocatalytic application, Nanoscale. (2014) 12740–12757.

[17] B.S. Shirke, P. V Korake, P.P. Hankare, S.R. Bamane, K.M. Garadkar, Synthesis and characterization of pure anatase TiO_2 nanoparticles, J Mater Sci Mater Electron. 22 (2011) 821–824.

[18] Y.Q. Jia, Crystal Radii and Effective Ionic Radii of the Rare Earth Ions, J. Solid State Chem. 187 (1991) 184–187.

[19] K.V. Kumar, K. Porkodi, F. Rocha, Langmuir-Hinshelwood kinetics - A theoretical study, Catal. Commun. 9 (2008) 82–84.

[20] S.M. Sheikh, K.E. O'Shea,G. Zhang, H.M. El-Hosainy, A.A. Ismail, P. Falaras, A.G. Kontos, D.D. Dionysious,, High performance sulfur, nitrogen and carbon doped mesoporous anatase-brookite TiO_2 photocatalyst for the removal of microcystin- LR under visible light irradiation, J. Haz. Mater. 280 (2014) 723-733.

[21] S. Hisaindee, M.A. Meetani, M.A. Rauf, Application of LC-MS to the analysis of advanced oxidation process (AOP) degradation of dye products and reaction mechanisms, TrAC - Trends Anal. Chem. 49 (2013) 31–44.

IOP Conf. Series: Materials Science and Engineering **195** (2017) 012007 doi:10.1088/1757-899X/195/1/012007

Synthesis, Structural and Morphological Property of $BaSnO_3$ Nanopowder Prepared by Solid State Ceramic Method

Jibi John[1], V.P Mahadevan Pillai[1]*, Anitta Rose Thomas[2], Reji Philip[2] Jaison Joseph[3], S.Muthunatesan[4], V.Ragavendran[5], Radhakrishna Prabhu[6]

[1]Department of Optoelectronics, University of Kerala,Thiruvanthapuram, Kerala,India
[2]Light and Matter Physics Group, Raman Research Institute, Bangalore, India
[3]Department of Physics, Government College, Khandola, Goa, India
[4] Department of Physics, Government Arts college, Kumbakonam, Tamil nadu, India
[5] Department of Physics, Sri Chandrasekharendra Saraswathi Viswa Mahavidyalaya University, Kanchipuram, Tamil nadu, India
[6]School of Engineering, Robert Gordon University, Aberdeen, United Kingdom

E-mail: vpmpillai9@gmail.com

Abstract. $BaSnO_3$ is a cubic perovskite-type oxide that behaves as an n-type semiconductor with a wide band gap of 3.4 eV and remains stable at temperatures up to 1000^0C. It has wide applications such as thermally stable capacitors, humidity sensors, gas sensors, etc. Barium stannate has also been used in optical applications, in capacitors and ceramic boundary layers, and as a promising material to produce gas phase sensors for the detection of carbon monoxide and carbon dioxide. $BaSnO_3$ powder was prepared by solid state ceramic method. X-ray diffraction pattern of the prepared sample presents all the characteristic peaks of cubic phase of $BaSnO_3$ (JCPDScard no: 15 -0780). The lattice constant for the compound was calculated and found to be $4.101A^0$ which is in agreement with the reported value ($4.112A^0$). The average size of the crystallites estimated by Debye Scherrer's formula was found to be 49 nm shows the nanostructured nature. The Raman bands observed ~ 139, 833 and 1122 cm^{-1} can be assigned on the basis of the fundamental vibrations of SnO_6 octahedron which has Oh symmetry, in the distorted perovskite structure. The SEM image shows a porous surface morphology with grains of cuboidal structure with well-defined grain boundaries. UV-Visible spectra shows $BaSnO_3$ powder exhibit high reflectance in the 400-700 nm range.

1. Introduction

Alkaline earth stannates with the general formula $RSnO_3$(R = Ba,Sr and Ca) are important material systems inview of their interesting physical properties and perovskite structures.[1] Perovskite-type oxides have a simple and flexible structure that is easy for ionic substitution, carrier doping and oxygen non-stoichiometry, which can form a vast set of technologically important materials for a wide variety of industrial applications[2,3]. $BaSnO_3$ is a cubic perovskite-type oxide that behaves as an n-type semiconductor with a wide band gap of 3.4 eV and remains stable at temperatures up to 1000^0C. It has wide applications such as thermally stable capacitors, humidity sensors, gas sensors, etc[4-5]. Barium stannate has also been used in optical applications,[6] in capacitors and ceramic boundary layers, [7]and as a promising material to produce gas phase sensors for the detection of carbon monoxide[8] and carbon dioxide.[9] Lampe et al[8] proposed thin films of $BaSnO_3$ to be used as a NO sensor in the temperature range of 450–550^{0}C, and as a material for CO sensor in the temperature range of 600–700^{0}C[10]. It is reported that $BaSnO_3$ doped with a few percent of La exhibits unusually high electrical mobility of 320 $cm^2(Vs)^{-1}$ at room temperature and superior thermal stability at high temperatures. Various techniques have been employed to synthesize, $BaSnO_3$ powder such as solid state method, self heat sustained method, co-precipitation method, hydrothermal synthesis method etc. This work reports the preparation and characterization of $BaSnO_3$ powder by solid state method and its structural, morphological and optical properties.

2. Experimental techniques

$BaSnO_3$ powder was prepared by solid state ceramic method[11]. Stoichiometric amount of commercially available high purity $BaCO_3$ and SnO_2 powders (Aldrich-purity 99.9%) were mixed in an agate motor for 6 hrs using acetone as the mixing medium. The homogeneous mixture thus obtained was heated in an alumina crucible up to temperature 1000°C and kept at that temperature for 8 hrs. Polyvinyl alcohol was added to the calcined powder for the fabrication of pellets, which was burnt out during high-temperature sintering. Circular disc shaped pellets were prepared by applying a load of 5 ton. These pellets were crushed and sintered in a platinum crucible at temperature 1250°C for 6 hrs. The structural, morphological and optical properties of the material were investigated in detail. The crystalline quality and crystallographic orientation were investigated using X-ray diffraction analysis (XPERT-PRO X-ray diffractometer) using Cu Kα1 radiation of wavelength 1.54060A° in the 2θ range 10-100°. The vibrational spectrum was recorded using micro-Raman spectrometer (Labram HR-800) using a laser radiation of wavelength 514 nm from an argon ion laser. The spectrum was recorded with a spectral resolution of 1cm^{-1}. The morphology and elemental analysis of the compound were investigated using NOVA NANOSEM 450 (FEI,U.S.A) equipped with Quantax 200 Energy Dispersive X-ray spectrometer (Bruker).The absorbance and reflectance spectra of the material in the spectral range 200-900 nm were recorded using JASCO V-550 UV-visible double beam spectrometer. The TEM observation of the sample was carried out by using JEOL JEM -2100 electron microscope operating at 200keV.

3. Results and discussion

3.1. XRD analysis and Micro-RAMAN analysis

The XRD pattern of $BaSnO_3$ powder prepared by solid state reaction method and it presents a polycrystalline nature. X-ray diffraction pattern of the prepared sample presents all the characteristic peaks of cubic phase of $BaSnO_3$ (JCPDS card no: 15 -0780). The inter planar distance 'd' of the powder is calculated using Bragg's relation 2dsinθ =nλ, where λ is the wavelength of the X-ray radiation and θ is the diffraction angle. The lattice constant for the compound was calculated and found to be 4.101A^0 which is in agreement with the reported value (4.112A^0)[1]. The average crystallite size (D_{hkl}) of the powder can be calculated using the Debye Scherrer's formula[12],

$$D_{hkl} = \frac{K\lambda}{\beta\cos\theta} \qquad \text{------- (1)}$$

where, λ is the wavelength of the X-ray and β is the FWHM in radians. The average size of the crystallites was found to be 49 nm. This shows the nanocrystalline nature of the powder.

Barium stannate in the cubic structure has space group Pm3m with 1 formula unit per Bravais lattice. The micro-Raman spectrum of the $BaSnO_3$ powder prepared in the present case by solid state reaction method presents a very intense band at 1054 , intense band at 560,139and weak intense band at 686,336,1130.The band at 1054 is due to$BaCO_3$. The Raman bands observed ~ 139, 833 and 1130 cm^{-1} can be assigned on the basis of the fundamental vibrations of SnO_6 octahedron which has O_h

IOP Conf. Series: Materials Science and Engineering **195** (2017) 012007 doi:10.1088/1757-899X/195/1/012007

symmetry, in the distorted perovskite structure. The Raman active mode observed at 139cm^{-1} can be assigned to v_5F_{2g} *mode*.[13]

3.2 SEM Analysis

Fig.1(a). shows the SEM micrograph of the $BaSnO_3$ sample. The SEM image shows a porous surface morphology with grains of cuboidal structure having well defined grain boundaries. The elemental analysis of the compound using EDS analysis supports the formation of the compound. Fig.1(b). shows the EDS spectra of the powder sample prepared by the solid state method. The EDS spectra also gives an evidence for the synthesis and formation of compound.

Fig.1(a):SEM images of $BaSnO_3$ prepared by solid state reaction method

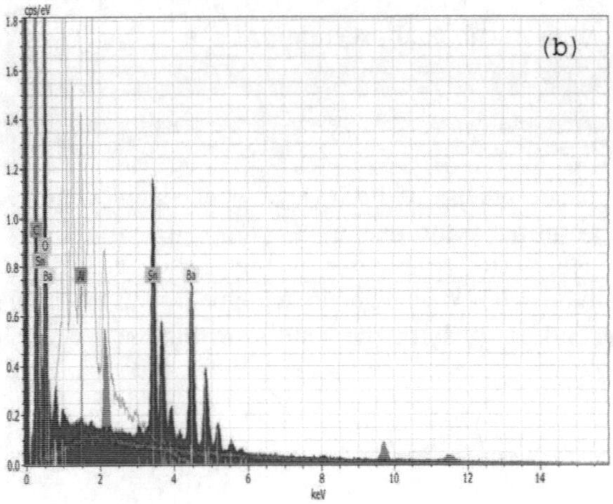

Fig. 1(b): EDS images of $BaSnO_3$ prepared by solid state reaction method

IOP Conf. Series: Materials Science and Engineering **195** (2017) 012007 doi:10.1088/1757-899X/195/1/012007

3.3. TEM Analysis

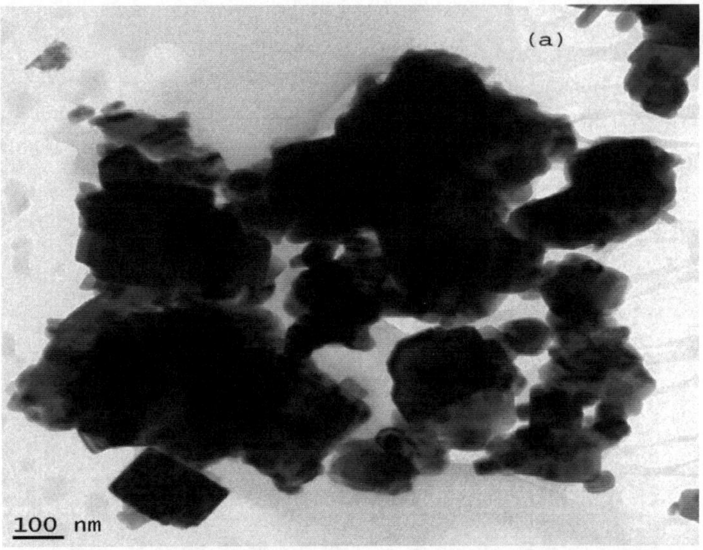

Fig 2: TEM image of $BaSnO_3$ prepared by solid state reaction method

A high resolution nanostructural and morphological characterization of powder were studied by using transmission electron microscopy The particle size of the powder calculated from *TEM analysis* is 55nm which is in agreement with XRD and SEM analysis. The TEM image also shows the porous surface morphology with grains of cuboidal structure.

3.4. UV-Visible Analysis

The reflectance and absorbance spectra of $BaSnO_3$ powder samples are recorded in the wavelength range 200-900 nm. The absorption spectrum and reflectance spectrum of the powder sample is shown in fig.3(a)and 3(b) respectively. The sample exhibit high reflectance in the 400-700 nm range. The band gap energy is found to be ~ 3.1eV which is in agreement with reported values. The band gap and the electronic structure depend primarily on the bond in within the corner-sharing octahedral network (SnO_6)[14].

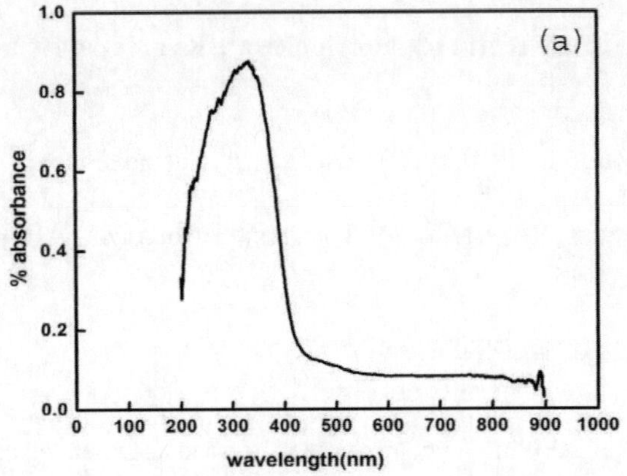

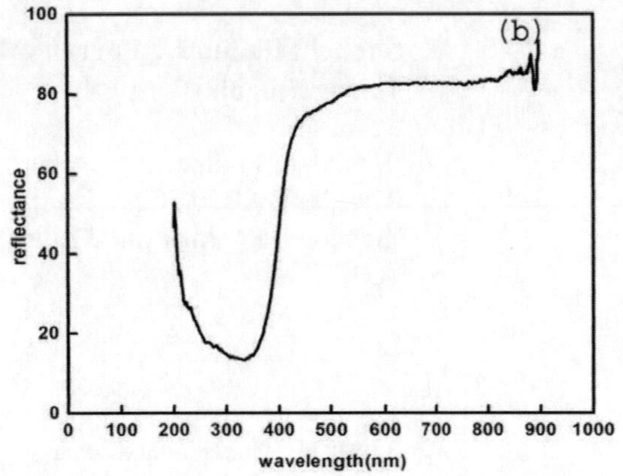

Fig.3(a):Absorption spectrum of $BaSnO_3$ and 3(b): Reflectance spectrum of $BaSnO_3$

4. Conclusion

The XRD and micro-Raman analysis suggest cubic crystalline structure supporting the formation of $BaSnO_3$ phase in the prepared sample. The SEM image shows a porous surface morphology with cuboidal grains with well- defined grain boundaries for the prepared sample. The sample exhibit high reflectance in the 400-700 nm range. The HRTEM image shows the nanocrystalline nature. The HRTEM image also supports the XRD and SEM analysis. The UV-visible analysis is carried out and the sample exhibit high reflectance in the 400-700 nm range. The band gap energy is found to be ~3.1eV which is in agreement with reported values.

5. References

[1] Qinzhuang Liu,etal 2010 *Journal of physics D: applied physics* 43 4554401
[2] Lampe U, Gerblinger J and Meixner H 1995 *Sensors & Actuators B* 26 97
[3] Tao S W, Gao F, Liu X Q and Sørensen O T 2000 *Sensors & Actuators B*—Chem. 71 223
[4].Cerda J, Arbiol J, Diaz R, Dezanneau G and Morante J R 2002 *Mater. Lett.* 56 131
[5]. Shimizu Y, Fukuyama Y, Narikiyo T, Arai H and Seiyama T 1985 *Chem. Lett.* 377
[6]. H. Mizoguchi, P. M. Woodward, C.-H. Park, and D. A. Keszler, 2004. *J. Am.Chem. Soc.*126, 9796.
[7] R. Vivekanandan and T. R. N. Knutty, 1988. *Ceram. Int.* 14, 207.
[8] U. Lampe, J. Gerblinger, and H. Meixner, 1995 *Sensors & Actuators*, B24/25, 657.
[9] T. Ishihara, K. Kometani, Y. Mizuhara, and Y. Takita, 1991. *Chem. Lett.*20,1711 .
[10] U. Lampe, J. Gerblinger, and H. Meixner, 1995 *Sensors & Actuators*, B26/27, 97
[11] S .Upadhyay, O. Parkash and D. Kumar, *1997 Journal of Mater. Science letters16 13300-1332*
[12] D.B Cullity, 1956 *Elements of X-ray diffraction* (Addison-Wesley Inc., Massachusetts)
[13] K.K. James, A. Aravind and M.K. Jayaraj, 2013 *Applied surface science* 282,121-125
[14] Shahram Soleimanpour, FaramarzKanjouri, 2014.*Physica* B4 32 16–20

IOP Conf. Series: Materials Science and Engineering **195** (2017) 012008 doi:10.1088/1757-899X/195/1/012008

Oil Based Drilling Fluid Waste: An Overview on Highly Reported But Less Explored Sink For Environmentally Persistent Pollutants

Shohel Siddique[1*], Lorraine Kwoffie[1], Kofi Addae-Afoakwa[1], Kyari Yates[2] and James Njuguna[1*]

[1]School of Engineering, Robert Gordon University, Riverside East, Garthdee Road, Aberdeen, AB10 7GJ, UK
[2]School of Pharmacy and Life Sciences, Robert Gordon University, Aberdeen, AB10 7GJ, UK

E: s.a.siddique@rgu.ac.uk , j.njuguna@rgu.ac.uk

Abstract. Operational discharges of spent drilling fluid, produced water, and accumulated drill cuttings from oil and gas industry are a continuous point source of environmental pollution. To meet the strict environmental standard for waste disposal, oil and gas industry is facing a numerous challenges in technological development to ensure a clean and safe environment. Oil and gas industry generates a large amount of spent drilling fluid, produced water, and drill cuttings, which are very different in every drilling operation in terms of composition and characterisation. This review article highlights the knowledge gap in identifying the different sources of waste streams in combined drilling waste. This paper also emphasises how different chemicals turn into environmentally significant pollutants after serving great performance in oil and gas drilling operations. For instance, oil based drilling fluid performs excellent in deeper drilling and drilling in harsh geological conditions, but ended with (produces) a significant amount of persistent toxic pollutants in the environment. This review paper provides an overview on the basic concepts of drilling fluids and their functions, sources and characterisation of drilling wastes, and highlights some environmentally significant elements including different minerals present in drilling waste stream.

1. Introduction

Oil and gas industrial exploration operational discharge, accidental spillage, or improperly disposed drilling wastes have serious detrimental effects on human and environment. When oil & gas drilling fluids and cuttings are disposed on the ground surface, the liquid fraction of chemicals start seeping through the ground and eventually these chemicals destroy the living organisms in the ground and pollute the groundwater [1]. Consequently, oil based drilling fluids (OBFs) and associated drill cuttings have become a major challenge in the industry for compliance with the requirements of zero discharge. Due to the European Union (EU) Waste Framework Directive (WFD), new legislations have been implemented in the EU member countries including United Kingdom for the recycling of wastes with a mission to prevent and reduce landfilling of waste [2,3].

In oil and gas industry, OBFs play an important role in drilling operation by providing relatively better shale inhibition, lubricity, and thermal stability characteristics [4,5]. It performs various functions including removing and cleaning drill cuttings from the drilling operation, cooling and lubricating the drill bit, controlling the hydraulic pressure, and ensuring a hydrostatic head to help integrating the wellbore wall and protecting from well blowouts [6,7]. Where reliable shale inhibition, better borehole stability, low filtration rates, low formation damage, low corrosion, and excellent lubricity are needed especially in sensitive well or geological conditions, invert-emulsion systems (OBFs and synthetic oil based fluid) are preferable [8,9]. The OBFs systems were developed and introduced in the 1960s to reduce and rectify different drilling issues such as, formation clays, swell and slough issues involved in WBFs, downhole temperature control, contaminants and stuck pipe and torque and drag [10,11,12]. In OBFs system, oil is the continuous phase and is formulated with diesel, mineral oil, or low-toxicity linear olefins and paraffins, (linear olefins and paraffins are sometimes considered as synthetic oils based on the synthesis from smaller molecules) [13]. Barium sulphate (barite) and organophillic bentonite are commonly used in the OBFs for increasing the system density and for improving the viscosity property respectively and therefore better functional properties [14].

Back in 1970, different environmental organisations and government agencies designed a guideline which has today been implemented in different processing industries including oil and gas industry [1]. It follows that drilling is one of the major chemical intensive operation in the oilfield and has a significant source of chemical exposure and subsequent health effects [15,16]. In oil based drilling operation, OBFs containing drill-cutting discharge in offshore-drilling is not permitted to most areas. During the drilling operation, a large amount of pollutants are introduced in drilling fluids, which is a big challenge with respect to solid waste handling in process operations. All such drill cuttings and spent drilling fluids are processed, and shipped to shore for disposal [17]. In case of onshore operations, pit burial is the most common technique used for drill cuttings management. In this process, drill cuttings have been temporarily stored in earthen pits for both onsite and offsite operation before disposal to the land or subsurface [18]. Hence, this review highlights the sources and characterisation of drilling wastes, and environmentally significant constituents present in drilling wastes to provide support for designing effective waste treatment plan.

2. Sources of drilling wastes

In oil and gas industry, the well drilling process produces two main types of wastes; drill cuttings, and spent mud [19]. In drilling operation, the pressure applied to penetrate the drill pipe into the oil and gas reservoirs produces a large amount of pieces of rock being drilled to fall to the bottom of the well bore [20,21]. These pieces, referred to as drill cuttings, clog the well if not carried out. Drilling fluid, also known as mud due to its consistency and appearance, is circulated in the well to transport the drill cuttings to surface [22,23]. In most deep drilling operation, oil based fluids (OBFs) are used instead of water based fluids (WBFs) depending on the drilling conditions for efficient and cost-effective operations. It is quite common practice to use both WBFs and OBFs in drilling the same well, with WBFs for shallow portion of the well and OBFs for deep drilling [19]. The sources of drilling wastes largely depend on the cuttings based on the geological condition of the borehole, the depth of the well, and the fluid used in the drilling operation [24]. The accumulation of drill cuttings is estimated in the range of 130 to 560 m^3 per well and the possible sources and components of wastes are presented below [25].

- Waste lubricants: lube oil, grease
- Spacers: mineral oil, surfactants
- Spent contaminated water based muds: mud, mineral oil
- WBM cuttings: formation solids, WBM, mineral oil
- Spent OBMs : mud, mineral oil
- OBM cuttings: formation solids, OBMs
- Spent bulk chemical: cement, bentonite, barites, viscosifiers, thinners, fluid loss protective additives
- Spent special products: scavengers, defoamers, tracers

The variations in drilling fluids composition and the variations in geological formation generate a complex mixture of drilling fluid wastes, which cannot be classified into any classical drilling fluid waste profile. For an example, in OBFs base oil may be of various natures and also the additives used in OBFs may be of very complex form such as surfactants, organophilic clays, viscosifyers [16,25]

3. Characteristics of drilling wastes

In 2004, World Oil's categorised drilling fluids into nine distinct types including dispersed freshwater, non-dispersed fresh water, saltwater, oil-based, synthetic-based, air, mist, foam, and gasified drilling fluid systems [10,12]. These drilling fluids can be broadly classified as either liquid or pneumatic [26]. Drilling fluid selection in a drilling operation depends mainly on the geological formation information of the wellbore area. However, drilling fluids should possess various physical properties, such as thixotropy and rheology to make the drilling operation economical and sustainable [27]. After the drilling operation, accumulated drill cuttings are suspended, assimilated, or dissolved in the drilling fluids without affecting its physical properties [28]. These fluids may contain a wide variety of dissolved minerals, dissolved and dispersed oil compounds, salts, metal ions, naturally occurring radioactive materials (NORM) and dissolved gases. To meet the environmental regulations, these fluids may need to be treated to a satisfactory level before disposing them in landfill. To identify the concerning constituents present in waste stream and to design the effective treatment process, the accurate and detailed physical and chemical characterisations of wastes are necessary [24].

The physical composition of drilling wastes is mainly based on the type of drill cuttings produced. These cuttings are generally a reflection of the geological constituents of the sub-surface being drilled as well as the individual solid or chemical components originally contained in the drilling fluid [19]. In 1996, drilling operation was estimated to accumulate 7 million m^3 drill cuttings in North Sea between the years 1964 and 1993 and was projected to 12 million m^3 by 2000. Although the sources and compositions of wastes vary from site to site, their behaviour towards biological activities, cohesion with oil rich silts remain the same in nature. In a typical drill cutting pile, the pile is assumed to compose of 20-60% water, a bulk density of 1.6-2.3 g/cm^3, and a particle size ranging from 10μm to 2 cm [29,30]. Hudgins [31] reported the most comprehensive study to date available in open literature covering ten operating companies and six chemical suppliers in North Sea that obtained data (see Fig 1) on the specific types and quantities of chemicals used in their operation and identified the properties of these chemicals. The survey [31] also presented the discharge quantities and concentrations of chemicals during exploration and production activities performed by these companies in the North Sea. However, based on the Hudgins [31] survey results it can be summarised that the WBM accumulated more than three times the volume of discharge compare to OBM. It also noticed that the weighting agents, salinity, and bentonitic chemicals accumulated about 90% of the total WBM discharge.

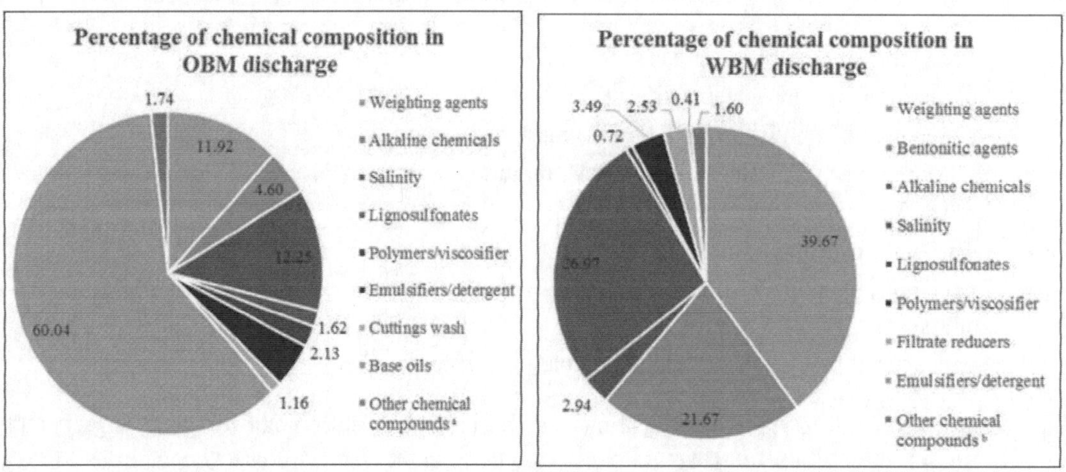

Fig. *1* Percentage of individual chemical constituents present in OBM and WBM discharge adapted from Hudgins [31].

Note: a) Bentonic agents, lost circulation, lignites, filtrate reducers, gilsonite, wetting agents, and thinning agents. b) Lost circulation, lignites, gilsonite, defoamers, biocides, corrosion inhibitors, scale inhibitors, drilling lubricants, pipe-release agents, oxygen scavengers, shale inhibitors, and thinning agents.

It can be concluded by highlighting that about 53% of chemicals that are used in drilling operation are discharged as wastes and thus cause the pollutants burden in the environment [31,32].

The chemical composition of drill cuttings at the time of disposal is an indication of the drilled sub-surface strata and concentration of the chemical components of the drilling fluid that remain attached to the cuttings. The temporal trends towards drilling activities provide important information on the long term effects of drill cutting discharges on geochemical and hydrogeological conditions [33]. Findings from different groups confirm the presence of certain metals in drill cuttings and their potential effect on the environment. Among these metals Cd, Cr, Ni, As, Co, Cu, Pb, V, Zn, Al, Ba, Fe, Mn are predominant in drill cuttings [34,35]. During the period of 1981-1986, the average annual discharge of oil on cuttings to the Norwegian Continental Shelf (NCS) was 1940 tons and that was eliminated gradually by implementing different directives such as OSPAR Commission in OSPAR regions. In addition to oil on cuttings discharge, the amount of produced water (PW) discharge has increased significantly due to the well ageing and also the rising number of oil and gas exploration and production fields. Large volume of OBM cuttings and SBM cuttings piled up in the seafloor before the regulations implemented in 1993/1996. It was estimated that about 45,000 m^3, a height of around 25m, and a footprint of more than 20,000 m^2 cutting piles are still present in the northern and central part of the North Sea. About 79 large (>5000 m^3) and 66 small (<5000 m^3) cutting piles have been identified in United Kingdom Continental Shelf (UKCS) and NCS. Further, a significant concentration of total hydrocarbons (10,000 to 600,000 mg/kg) exist in the North Sea piles [36].
However even WBM cuttings with less hydrocarbon content may seriously affect benthic fauna by elevating oxygen consumption in sediments. The risk of drilling waste associated with WBM cuttings discharge to the ecosystem is presently considered low, but this statement cannot be verified from the published literature. However, the variation in the amount of drilling fluid that remains adhered to the drill cuttings at the time of disposal is influenced by the size of the cuttings. The smaller the size of cuttings, the harder it is to separate it from the drilling fluid. In addition to these metals derived from the drilling fluid and geologic formation being drilled, cuttings may also contain some petroleum hydrocarbons closely linked to that of the reservoir rock [33].

4. Environmentally concerned constituents in drilling waste

Soil contamination is hazardous to health and environment through its action on surface waters, ground waters and vegetation (phytotoxicity, bioaccumulation). Oil and gas industries, like other process industries, have a detrimental effect on environment [37]. The hazardous effect of the environmentally significant constituents in the produced drilling wastes is predominantly dependent on each constituent, its concentration at exposure, biotic environment at point of discharge and the duration of exposure.

The typical type of drilling wastes and their potential constituents are [25]:

- WBM cuttings: Heavy metals, inorganic salts, biocides, hydrocarbons
- OBM cuttings: Heavy metals, inorganic salts, hydrocarbons, solid/cuttings
- Spent OBM: Heavy metals, inorganic salts, hydrocarbons, solid/cuttings, BOD, surfactants
- Spent WBM: Metals including heavy metals, inorganic salts, hydrocarbons, biocides, hydrocarbons, solid/cuttings, BOD,
- Waste lubricants: Heavy metals, organic compounds.

Some of the metals are suspected to be present in concentrations significantly higher than the naturally occurring concentrations of the sediments, which makes the disposal of these wastes a critical environmental concern [25,34]. Arsenic, nickel, copper, chromium, zinc, anthracene, diuron, fluoranthene, naphthalene, phenanthrene, and pyrene are considered as environmentally significant chemicals according to the literature [35,37,38].

Although the amount of drilling fluid constituents is very low, most of them especially heavy metals have a chronic effect on environment. The pollutants may nowadays be categorised in two different groups of pollutants: List I and List II as shown on Table 1 according to European Council Directive 76/464/EEC [39]. List I group are substances, which are toxic, persistent, and possess the bioaccumulation properties while List II is a group of chemicals, which have deleterious effect on the aquatic environment. However, list II pollutants can be confined to a given area and also the pollutants concentration varies based on the characteristics and location of the water into which the pollutants are discharged.

Table 1: List I and II pollutants in environment.

Type of pollutants	Members of pollutant groups
	Organohalogen compounds and substances
	Organophosphorus compounds
	Organotin compounds
List I	Carcinogenic substances
	Mercury and its compounds*
	Cadmium and its compounds*
	Persistent mineral oils and hydrocarbons of petroleum origin
	Persistent synthetic substances
	Certain metals, metalloids, and their compounds: 1) Zinc 2) Copper* 3) Nickel* 4) Chromium (Cr(VI)*)
	5) Lead* 6) Selenium* 7) Arsenic* 8) Antimony* 9) Molybdenum 10) Titanium 11) Tin* 12) Barium
	13) Beryllium 14) Boron 15) Uranium 16) Vanadium 17) Cobalt 18) Thalium* 19) Tellurium* 20) Silver
	Biocides and their derivatives
List II	Toxic or persistent organic compounds of silicon and its substances
	Inorganic compounds of phosphorus and elemental phosphorus
	Non persistent mineral oils and hydrocarbons of petroleum origin
	Cyanides and fluorides
	Substances causing oxygen imbalance such as ammonia, nitrites

*: Hazardous waste classified in according to Directive 2008/98/EC

In 2008, the Waste Framework Directive 2008/98/EC brought legal changes to the list of waste and hazardous waste criteria based on the source and composition of wastes [40]. In this amendment the source of waste is identified into 20 chapters (from 01 to 20) and the different types of waste in the list are fully defined by the six-digit code (first two digits is chapter heading and the rest four digits for identifying sub-groups). Based on this Commission Decision, environmentally significant and hazardous elements or compounds which are present in drilling fluid wastes are denoted by * mark in Table 1 [41].

In addition, country specific requirements for discharge of drilling fluids and cuttings also do play a role in environmental protection. For instance, in the United Kingdom the OSPAR2000/3 discharge regulation comes into play and compliance requires limit of less than 1% oil on cuttings and do advise on when to inject cuttings or return to shore and oil recovery. The OSPAR2000/3 regulation does not permit discharge of synthetic based mud (SBM) cuttings offshore [39,42].

5. Future trends on drilling waste

To meet the strict environmental regulations, sustainable and effective waste management remains a big challenge in oil and gas industry. Fortunately, new waste treatment or clean-up operation may eliminate this problem and in addition, we should explore a new window to turn these hazardous wastes into value added products [44]. To utilise these pollutant materials which exist in wastes, it is very important to understand the sources of drilling fluid wastes, chemical composition, and characterisation of these wastes. Since oil based drilling fluids (OBFs) consist of diesel or mineral oil containing different types of polycyclic aromatic hydrocarbons (PAHs) and are also considered as a flammable hazard source, care consideration should be taken to design the cleaning or treatment processes [45].

Different mechanisms have been developed and continue to improve aimed at treating drill cutting waste including non-biological treatment processes and disposal options and bioremediation technologies for treating drill cuttings [20,2]. However, the potential environmental impacts of spent drilling fluids and drill cuttings after treatment are still considered as serious health and safety concerning issues [46,47,48].

Today, significant amount of drilling wastes accumulated during drilling operations are disposed off in the landfill or seabed without recovering the useful elements/compounds present in these wastes [18,49]. Treating these wastes to reuse and recycle them in different beneficial uses remains a significant challenge and any step to improve these processes are considered as sustainable and effective measures to reduce the environmental pollution in future [48,49,50,51]. For instance, the Waste to Want research being run at Centre for Advanced Engineering Materials at Robert Gordon University on the novel application of nanoclays extracted from spent oil based drilling mud (drilling fluid) clean-up as nanofiller in the manufacture of nanocomposite materials offers new solutions [44,52]. To use the beneficiary elements or compounds present in drilling fluid waste, it is important to first comprehensively analyse the composition and characterisation of this waste.

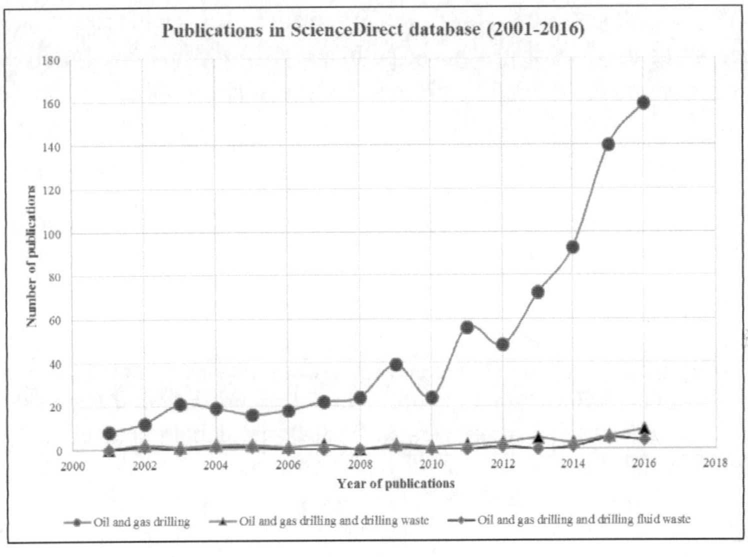

Fig. 2: Articles published in ScienceDirect database from 2001-2016

Analysis of recent publication records since 2001 (Fig 2) shows a continuous positive trend on research in oil and gas drilling operation though disappointing little to show on oil and gas drilling waste and drilling fluid waste. The survey covers the key words search in title, abstract, and keywords in publications from 2001 to 2016. 771 articles have been published during this time based on oil and gas drilling whereas only 41 papers have been published on oil and gas drilling waste. The number is even fewer in drilling fluid waste related publications which is 15 publications in last 16 years. There is a need of extensive research on drilling waste, especially drilling fluid related waste which meets cost savings needs and increased environmental awareness.

This paper highlights that composition of drilling fluid waste and the importance of making this waste into resources is not widely explored. Drilling fluid waste accumulation by its individual components might be a potential area where more research work needs to focus to optimise the use of individual drilling waste constituents in reuse or recycling operation. Fig 2 depicts the increasing interest in drilling operation related publications, but ironically the interest in investigation to the effect of individual ingredients of drilling wastes on environment is not well documented which is needed for a sustainable waste management in oil and gas exploration and production industry.

6. Conclusion

The accumulated drilling fluid wastes in oil and gas industry is different in every operation site. The variations in drilling operations including using drilling fluid with different compositions and the variations in geological conditions make these waste streams so diverse that there is not any standard drilling fluid waste profile tool to identify the composition and character of the wastes. The scenario is even more complex in offshore drilling operation as the unique sediment characteristics, benthic community, and hydrodynamic regime also influence the drilling waste characteristics.

To design a sustainable and viable drilling waste management plan/model, the first step is to identify the composition and nature of the pollutants in the wastes. Based on this information different waste treatment plan can be placed in operation such as, thermal treatment, thermo-mechanical treatment, biological treatment, encapsulation of pollutants. To meet the current strict environmental regulations with respect to disposal of this waste or to recycle or recover valuable components such as certain metals, identification of drilling waste composition and characterisation analysis is the obvious first step to move forward to the next stages. However, the environment is still facing the biggest challenge - are these treatment measures capable of producing pollutants free or environmentally safe discharge or producing pollutants free or eco-friendly solid waste? The solution might be very difficult as these treatment processes involve space requirements, duration of treatment operation, operational cost, investment cost, monitoring requirement, expertise etc. These obligations open up a new era of research to use this waste as a raw material to make valuable products rather than incurring burden to the environment.

References

[1] Caenn R, Darley HCH, Gray GR. Chapter 12 - Drilling and Drilling Fluids Waste Management. In: Caenn R, , Darley HCH, , Gray GR, editors. *Composition and Properties of Drilling and Completion Fluids* (Sixth Edition) Boston: Gulf Professional Publishing; 2011. p. 617-654.

[2] Managing Onshore Drilling Wastes-Abu Dhabi Experience. Abu Dhabi International Petroleum Exhibition and Conference: Society of Petroleum Engineers; 2000.

[3] Fijał J, Gonet A, Jamrozik A. Characterization, properties and microstructure of spent drilling mud from the point of view of environmental protection. *AGH Drilling, Oil, Gas* 2015;**32**(3):483--492.

[4] Sönmez A, Verşan Kök M, Özel R. Performance analysis of drilling fluid liquid lubricants. *Journal of Petroleum Science and Engineering* 2013 **8**;108:64-73.

[5] Kania D, Yunus R, Omar R, Abdul Rashid S, Mohamad Jan B. A review of biolubricants in drilling fluids: Recent research, performance, and applications. *Journal of Petroleum Science and Engineering* 2015 **11**;135:177-184.

[6] Drilling fluid: State of the art and future trend. North Africa Technical Conference and Exhibition: Society of Petroleum Engineers; 2012.

[7] Abimbola M, Khan F, Khakzad N. Dynamic safety risk analysis of offshore drilling. *J Loss Prev Process Ind* 2014;**30**:74-85.

[8] Ismail AR, Aftab A, Ibupoto ZH, Zolkifile N. The novel approach for the enhancement of rheological properties of water-based drilling fluids by using multi-walled carbon nanotube, nanosilica and glass beads. *Journal of Petroleum Science and Engineering* 2016 3;**139**:264-275.

[9] Steinsvåg K, Bråtveit M, Moen B, Li V, Austgulen T, Hollund BE, et al. Expert assessment of exposure to carcinogens in Norway's offshore petroleum industry. *Journal of Exposure Science and Environmental Epidemiology* 2008;**18(2)**:175-182.

[10] SPE International. Drilling fluid types. 2015; Available at: http://petrowiki.org/Drilling_fluid_types. Accessed 04/14/2016.

[11] Silicate-based drilling fluids: competent, cost-effective and benign solutions to wellbore stability problems. SPE/IADC Drilling Conference: Society of Petroleum Engineers; 1996.

[12] Chapter II - Drilling Fluids. In: Fink JK, editor. Water-Based Chemicals and Technology for Drilling, Completion, and Workover Fluids Boston: Gulf Professional Publishing; 2015. p. 5-114.

[13] Chapter 1 - Drilling muds. In: Fink J, editor. Petroleum Engineer's Guide to Oil Field Chemicals and Fluids (Second Edition) Boston: Gulf Professional Publishing; 2015. p. 1-61.

[14] Arthur G. Clem and Robert W. Doehler INDUSTRIAL APPLICATIONS OF BENTONITE. American Colloid Company, Skokie, Illinois.

[15] Force DFT. Drilling fluids and health risk management.A guide for drilling personnel, managers and health professionals in the oil and gas industry 2009.

[16] Coussot P, Bertrand F, Herzhaft B. Rheological behavior of drilling muds, characterization using MRI visualization. *Oil & gas science and technology* 2004;**59(1)**:23-29.

[17] Hou B, Liang C, Deng H, Xie S, Chen M, Wang R. Oil removing technology of residues from waste oil-based drilling fluid treated by solid-liquid separation. *Journal of Residuals Science & Technology* 2012;**9**(4).

[18] Ball AS, Stewart RJ, Schliephake K. A review of the current options for the treatment and safe disposal of drill cuttings. *Waste Manag Res* 2012 May;**30(5)**:457-473.

[19] Offshore discharge of drilling fluids and cuttings-a scientific perspective on public policy. Rio Oil and Gas Conference. Rio de Janeiro, Brazil; 2000.

[20] Drilling waste management: past, present, and future. SPE Annual Technical Conference and Exhibition: Society of Petroleum Engineers; 2002.

[21] Carignan MP, Lake CB, Menzies T. Assessment of two thermally treated drill mud wastes for landfill containment applications. *Waste Manag Res* 2007 Oct;**25(5)**:394-401.

[22] Magalhães S, Scheid C, Calçada L, Lutterbach L, Rezende R, Waldmann A. Real time prediction of suspended solids in drilling fluids. *Journal of Natural Gas Science and Engineering* 2016;**30**:164-175.

[23] Quintero, Lirio, Antonio Enrique Cardenas, and David E. Clark. "Nanofluids and methods of use for drilling and completion fluids." U.S. Patent No. 8,822,386. 2 Sep. 2014.

[24] Piszcz K, Luczak J, Hupka J. Mobility of shale drill cuttings constituents. Physicochemical Problems of Mineral Processing 2014;50.

[25] Onwukwe S, Nwakaudu M. Drilling wastes generation and management approach. International *Journal of Environmental Science and Development* 2012;**3(3)**:252.

[26] Azar JJ, Samuel G.R. Drilling engineering. : *PennWell* Books; 2007.

[27] Besq A, Malfoy C, Pantet A, Monnet P, Righi D. Physicochemical characterisation and flow properties of some bentonite muds. *Appl Clay Sci* 2003;**23**(5):275-286.

[28] Zhou D, Zhang Z, Tang J, Wang F, Liao L. Applied properties of oil-based drilling fluids with montmorillonites modified by cationic and anionic surfactants. *Appl Clay Sci* 2016;**121**:1-8.

[29] Breuer E, Stevenson A, Howe J, Carroll J, Shimmield G. Drill cutting accumulations in the Northern and Central North Sea: a review of environmental interactions and chemical fate. *Mar Pollut Bull* 2004;**48(1)**:12-25.

[30] Breuer E, Shimmield G, Peppe O. Assessment of metal concentrations found within a North Sea drill cuttings pile. *Mar Pollut Bull* 2008;**56(7)**:1310-1322.

[31] Hudgins Jr CM. Chemical use in North Sea oil and gas E&P. *J Pet Technol* 1994;**46(01)**:67-74.

[32] Marsh R. A database of archived drilling records of the drill cuttings piles at the North West Hutton oil platform. *Mar Pollut Bull* 2003;**46(5)**:587-593.

[33] Phillips C, Evans J, Hom W, Clayton J. Long-term changes in sediment barium inventories associated with drilling-related discharges in the Santa Maria Basin, California, USA. *Environmental toxicology and chemistry* 1998;**17(9)**:1653-1661.

[34] Grant A, Briggs AD. Toxicity of sediments from around a North Sea oil platform: are metals or hydrocarbons responsible for ecological impacts? *Mar Environ Res* 2002 2;**53(1)**:95-116.

[35] Pozebon D, Lima EC, Maia SM, Fachel JMG. Heavy metals contribution of non-aqueous fluids used in offshore oil drilling. *Fuel* 2005 1;**84(1)**:53-61.

[36] Bakke T, Klungsøyr J, Sanni S. Environmental impacts of produced water and drilling waste discharges from the Norwegian offshore petroleum industry. *Marine environmental research.* 2013; **92**:154-169.

[37] Khodja M, a BergayaFa, Canselier JP, Khodja-Saber M, Cohaut N. Drilling fluid technology: performances and environmental considerations. : INTECH Open Access Publisher; 2010.

[38] Bignert A, Cossa D, Emmerson R, Fryer R, Füll C, Fumega J, et al. OSPAR/ICES workshop on the evaluation and update of background reference concentrations (B/RCs) and ecotoxicological assessment criteria (EACs) and how these assessment tools should be used in assessing contaminants in water, sediment, and biota. 2004.

[39] DIRECTIVE HAT. Council Directive 76/464/EEC of 4 May 1976 on pollution caused by certain dangerous substances discharged into the aquatic environment of the Community. *Official Journal L* 1976;**129(18/05)**:0023-0029.

[40] Parliament E. Directive 2008/98/EC of the European Parliament and of the Council on waste and repealing certain Directives. *Official Journal of the European Union, L.* 2008; 312:3-30.

[41] EUROPEANCOMMISSION. COMMISSION DECISION of 18 December 2014 amending Decision 2000/532/EC on the list of waste pursuant to Directive 2008/98/EC of the European Parliament and of the Council . EUROPEAN COMMISSION; 2014.

[42] OSPAR Commission. Assessment of the discharges, spills and emissions to air on the Norwegian Continental Shelf, 2009-2013 United Kingdom: OSPAR Commission; 2015.

[43] Bernier R, Garland E, Glickman A, Jones F, Mairs H, Melton R, *et al.* Environmental aspects of the use and disposal of non aqueous drilling fluids associated with offshore oil & gas operations. International Association of Oil & Gas Producers Report 2003;342.

[44] Waste to Want: Polymer nanocomposites using nanoclays extracted from Oil based drilling mud waste. IOP Conference Series: Materials Science and Engineering: IOP Publishing; 2014.

[45] Xie S, Jiang G, Chen M, Li Z, Mao H, Zhang M, et al. Treatment Technology for Waste Drilling Fluids in Environmental Sensitivity Areas. Energy Sources, Part A: *Recovery, Utilization, and Environmental Effects* 2015;**37(8)**:817-824.

[46] Rozell DJ, Reaven SJ. Water pollution risk associated with natural gas extraction from the Marcellus Shale. *Risk Analysis* 2012;**32(8)**:1382-1393.

[47] Mor S, Ravindra K, Dahiya R, Chandra A. Leachate characterization and assessment of groundwater pollution near municipal solid waste landfill site. *Environ Monit Assess* 2006;**118**(1-3):435-456.

[48] Shon C, Estakhri CK, Lee D, Zhang D. Evaluating feasibility of modified drilling waste materials in flexible base course construction. *Constr Build Mater* 2016;**116**:79-86.

[49] Xie S, Jiang G, Chen M, Li Z, Huang X, Liang C, et al. Harmless Treatment Technology of Waste Oil-Based Drilling Fluids. *Petrol Sci Technol* 2014;**32(9)**:1043-1049.

[50] Malachosky, Edward, and Donna M. Lantero. "Disposal and reclamation of drilling wastes." U.S. Patent No. 4,942,929. 24 Jul. 1990.

[51] Drilling waste management: past, present, and future. SPE Annual Technical Conference and Exhibition: Society of Petroleum Engineers; 2002.

[52] Njuguna J, Adegbotolu UV, Matthews K, Yates K. A study on polyamide nanocomposite using nanoclays nanofillers reclaimed from oil and drilling fluids and cuttings waste. 4th International Symposium on Energy, Challenges & Mechanics, 2015.

IOP Conf. Series: Materials Science and Engineering **195** (2017) 012009 doi:10.1088/1757-899X/195/1/012009

Tensile and Flexural Properties of Hybrid Graphene Oxide / Epoxy Carbon Fibre Reinforced Composites

G Watson[1], K Starost[1*], P Bari[2], N Faisal[1], S Mishra[2] and J Njuguna[*1]

[1] Centre for Advanced Engineering Materials, School of Engineering, Robert Gordon University, Aberdeen, AB10 7GJ, United Kingdom

[2] University Institute of Chemical Technology, North Maharashtra University, Jalgaon-425001, Maharashtra, India

E-mail: k.starost@rgu.ac.uk, j.njuguna@rgu.ac.uk

Abstract. In this study, nano-sized graphene oxide sheets were homogenously dispersed via sonication methods in epoxy followed by vacuum resin infusion for the fabrication of the epoxy, graphene oxide (GO) and micro-sized carbon fibre reinforced nanocomposites (EP/CF/GO). Graphene oxide concentrations ranging from 0.1 – 0.5 wt. % were studied to investigate the effect on tensile and flexural strength. It was observed that the tensile strength of the EP/CF decreased with the addition of GO but increased with GO weight concentration in the nanocomposites studied from 498MPa to 519 MPa for the inclusion of 0.1 to 0.5 wt.% GO respectively. The 0.5 wt. % EP/CF/GO recorded a 10% increase in Young's modulus compared to the classical epoxy / carbon fibre composites, and similar trend was observed for the flexural properties. However flexural strength of the GO samples did not surpass the control sample (epoxy /carbon fibre composites) with the 0.3 wt.% GO samples (EP/CF/GO) providing the greatest flexural strength of >580 MPa compared to the 0.1 wt.% and 0.5 wt.% GO samples.

1. Introduction

The development in carbon based fillers since the discovery of graphene has demonstrated significant improvement over the past decade. This has triggered interest and increasing implementation within commercial applications for both textile and engineering hybrid composites [1-4]. Use of reduced graphene and graphene oxide (GO) in epoxy-carbon fibre reinforced composites is still at infancy stages has been limited however due to the challenges in processing and dispersion of the fillers along with the high price associated therewith [5]. Homogeneous dispersion of the filler within the polymer and the strong interfacial interactions required between the filler and the matrix are the two biggest concerns when fabricating polymer nanocomposites [6].

The oxygen groups within GO offer and allow for a versatile, less fastidious and enhanced chemical cross-interlocking with the polymer chains [7]. The oxygen functional groups that GO possesses on its basal planes and edges permit it to be manipulated, exfoliated and functionalised to yield well-dispersed solutions of graphene oxide sheets [8-10]. Various studies have already presented GO to improve the polymer matrix material mechanical properties [11-13], and have struggled so far to achieve a superior dimensional performance and relative cost benefit to micro sized carbon fibre composites (CFCs) [14]. Since the inclusion of GO in epoxy has presented improved mechanical and electrical properties with reference to the unreinforced epoxy, the filler will potentially carry improvements in a fibre composite depending on the interfacial interaction. Researchers have therefore started to investigate combining GO nanofillers with the micro sized carbon fibre fillers [15-17]. This aim of this study is to experimentally investigate the combination of established carbon fibre reinforcement with graphene oxide to further enhance the hybrid material mechanical properties.

2. Methodology

2.1. Materials and Samples Manufacturing

A commercially available high performance bisphenol-A-(epichlorhydrin) based epoxy resin specifically formulated for use in vacuum resin infusion from Easycomposites (IN2 Epoxy Infusion Resin[@]) combined with a polyoxypropylendiamin based hardener from Easycomposites (AT30 Epoxy Hardener – Slow[@]) was chosen for the matrix. Graphene oxide (GO) flakes, 15-20 sheets with 4-10% edge-oxidized from Sigma-Aldrich (796034 Aldrich) was employed in this investigation. The 3k 2/2 Twill Carbon Fibre was obtained from Easycomposites.

The composite samples were manufactured through the vacuum resin infusion method. Concentrations of 0.1 wt. %, 0.3 wt. % and 0.5 wt. % were initially dispersed within methanol with the use of a sonication bath for 1 hour to allow for later dispersion of the GO in the Epoxy. Once fully dispersed, the solution was then homogenously dispersed within the bisphenol-A-(epichlorhydrin) based epoxy together with the hardener using a magnetic stirrer and manual mixing to allow for slow solvent evaporation. This was followed by the vacuum resin infusion process with 6 layers of the carbon fibre textile layered within a mould and left to cure. A reference sample without any GO was also manufactured, that is, EP/CF, EP/CF/GO with 0.1 wt. % GO, EP/CF/GO with 0.3 wt. % GO and EP/CF/GO with 0.5 wt. % GO.

2.2. Mechanical Testing

The manufactured samples were machined to 250x25mm for the tensile test and 100x15mm for the flexural test in accordance to the respective standard regulation ISO 527-4 tensile test [18] and ISO 178 3-point bend test [19]. The samples underwent a 3-point bend flexural test in accordance to ISO 178 at 2 mm/min and tensile test in accordance to ISO 527-4 at 1 mm/min [18-19]. We used a Instron Model 3382 universal testing machine to carry out the tensile and flexural testing.

3. Results and Discussion

3.1. Tensile Properties

The results from the tensile test conducted on the EP/CF composites with different wt. % of GO is displayed in Figure 1. The data represented is an average from the test samples recorded from the repeated tests. The concentrations of GO can be seen to have little or contrasting effects on the EP/CF based composite performance. Figure 1 indicates the non-reinforced EP/CFC sample to have the highest tensile strength. However, the Young Modulus can be seen to improve for the 0.3 wt. % and 0.5 wt. % concentrations of GO relative to the EP/CF sample. A comparison of the ultimate tensile strength (UTS) and Young's modulus is illustrated on Figure 2.

The increase in concentration of GO observed a decreasing effect on the extension of the specimen at the point of failure. The EP/CF ruptured after an extension of 5.35mm in comparison with the EP/CF 0.1 wt.%, 0.3 wt.% and 0.5 wt.% EP/CF/GO samples breaking at 4.73mm, 4.57mm and 4.51mm respectively. The materials are therefore appearing to become slightly more brittle with increase in GO concentration. This is shown in the increase of Young's Modulus results illustrated in Figure 2 for the 0.3 wt.% and 0.5 wt.% EP/CF/GO reinforced samples.

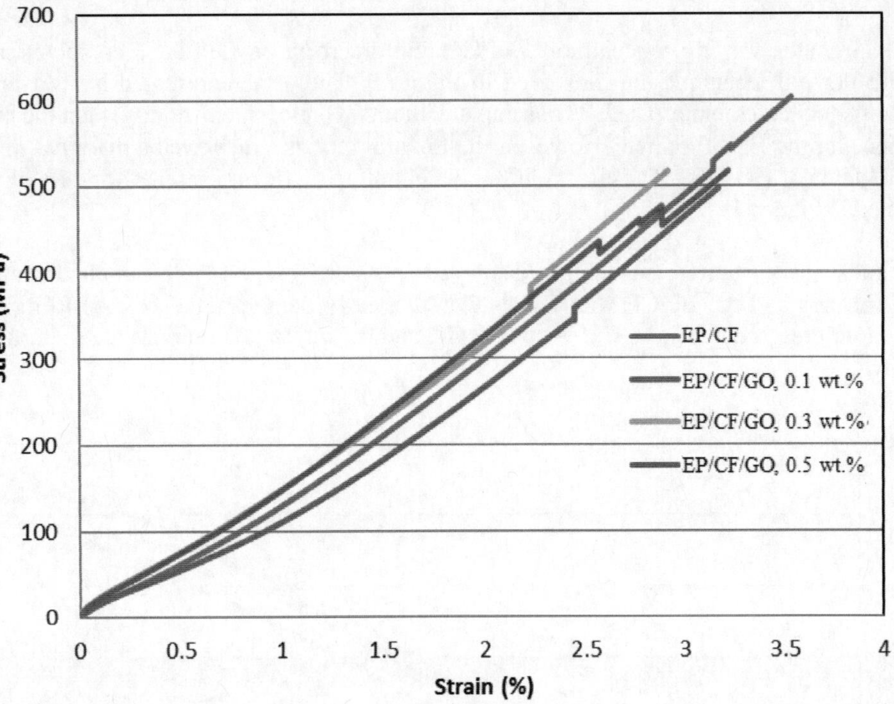

Figure 1: Stress vs strain from tensile test carried out on EP/CF based composites with varying GO concentrations

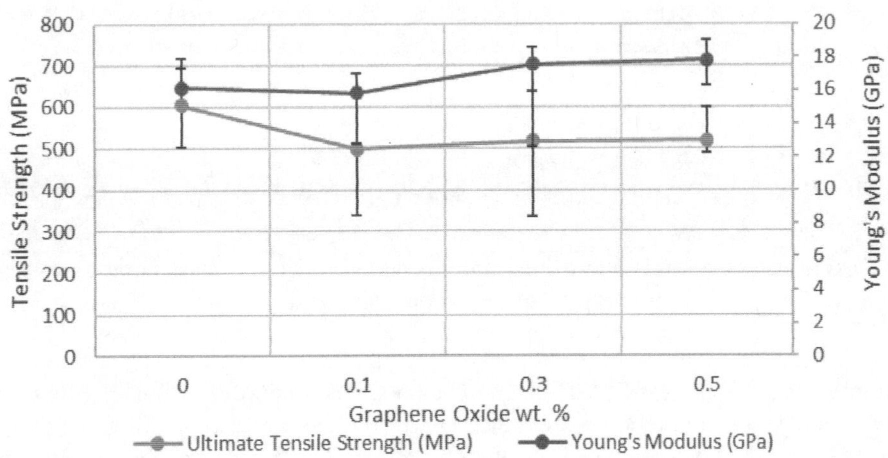

Figure 2: Tensile strength and Young's modulus of EP/CF based composites with varying GO wt.% concentrations

Figure 1 and Figure 2 both demonstrate the minimal effect the GO concentrations appeared to have on the EP/CF based material performance. Furthermore, the significant error bars indicate a high variety in the specimen performance. There was therefore a limitation with the material fabrication consistency. Further analysis is required to analyse the samples and the fabrication process. A possible reason could be due to poor or inconsistent bonding between the GO and CF, which is subject to further investigations. The slight increase of 11% in Young's modulus and UTS increase from 498.2MPa for the 0.1wt.% sample to 518MPa for the 0.5wt.% sample clearly demonstrates a potential material augmenting tensile strength performance.

IOP Conf. Series: Materials Science and Engineering **195** (2017) 012009 doi:10.1088/1757-899X/195/1/012009

3.2. Flexural Properties

The results from the varying concentrations of GO reinforcement on the EP/CF composite from the 3-point bend flexural technique are illustrated in Figure 3. The data represented is an average of the repeated tests for each sample. The EP/CF sample without GO reinforcement displayed the best flexural performance for the EP/CF sample. However, the flexural modulus displayed a marginal improvement with the addition of GO filler. The effect of GO concentration on flexural strength and flexural modulus is illustrated in Figure 4.

The flexural stress vs strain relationship is displayed on Figure 3. The EP/CF sample demonstrated the best performance and the EP/CF with 0.1 wt.% GO clearly demonstrated best performance for the grapheme reinforced composites. The 0.3 wt. % GO and 0.5 wt. % GO samples are seen to only differ marginally.

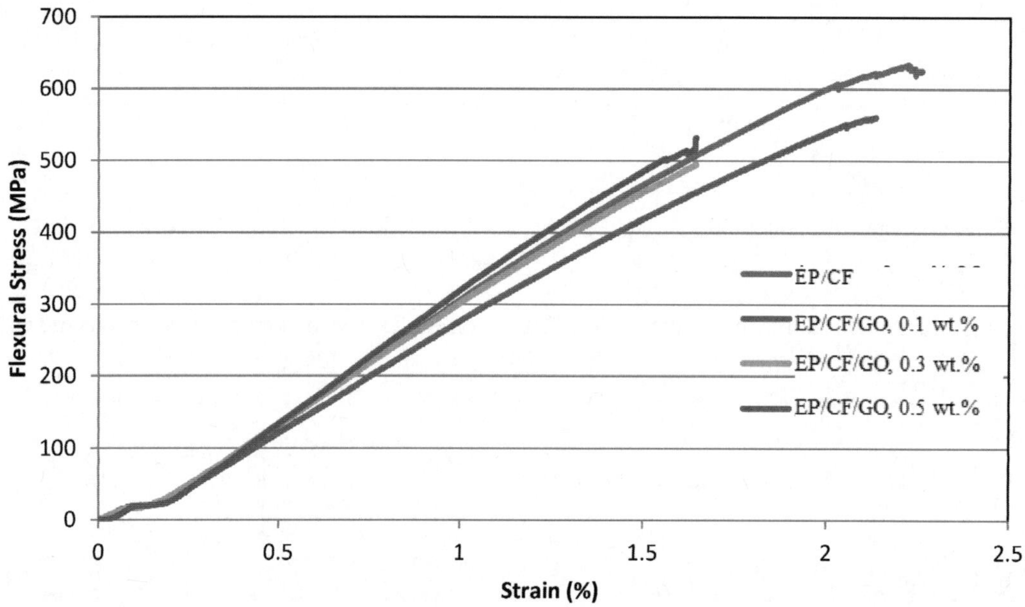

Figure 3: Flexural stress vs strain plot from 3-point bend test of EP/CF based composites materials with varying GO concentrations

As shown in Figure 4, the addition of 0.1 wt. % of GO in the EP/CF composite presented a decrease of 7% in flexural strength from neat EP/CF sample from 602MPa to 560MPa. The slight increase of 0.3 wt. % GO displayed an increase of 4% from the 0.1 wt. % GO sample to 582MPa, which is however still a decrease from the EP/CF sample. The then 9% decrease to 0.5 wt.% GO sample would indicate a performance threshold in GO concentration. Although the flexural strength between the GO reinforced samples displayed a slight increase, no sample improved performance from the EP/CF sample.

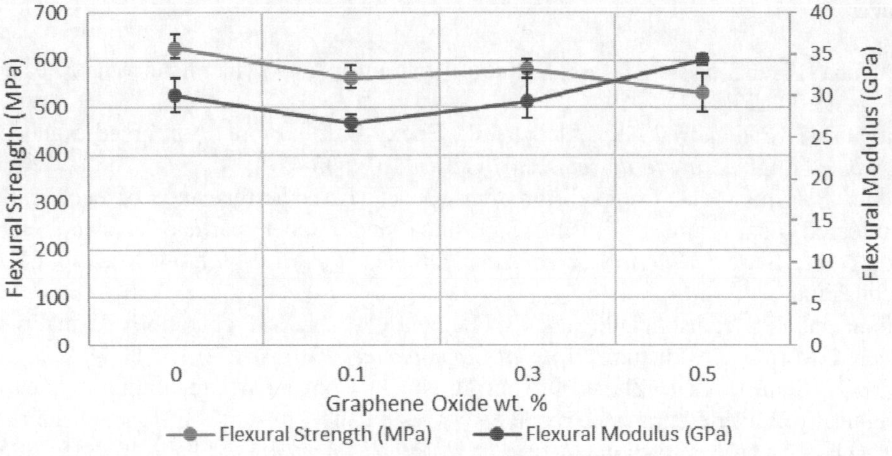

Figure 4: Flexural strength and modulus of EP/CF based composites with varying GO wt.% concentrations

However, as shown in Figure 3 and Figure 4, the use of GO as a filler displayed an increase in flexural modulus. Initially the 0.1 wt. % GO sample displayed a decrease in bending modulus from the EP/CF sample, but the 0.3 wt. % GO and 0.5 wt. % GO samples revealed a 10% and 30% increase from the 0.1 wt. % GO sample respectively. The 0.5 wt. % GO sample with a flexural modulus of 34.3GPa demonstrated a 14% increase from the reference EP/CF sample. The increase in stiffness however displayed a decrease in overall flexural strength from the EP/CF sample.

This is part of an ongoing study. The analysis will further comment on microstructures on broken surfaces, microscopy on samples surface, thermal and electrical conductivity all due to the varied concentration dispersions of the GO on epoxy samples.

4. Conclusion

The initial findings from this study have appeared to show little effect from the incorporation of GO in a EP/CF composite. In comparison to the reference EP/CF material, the GO samples did not show any improvement in tensile strength or flexural strength. However, a minor improvement in Young's modulus and flexural strength were observed with the addition of the GO into the EP/CF sample. The increase in GO concentration did have a marginal effect on tensile and flexural strength when solely comparing the different GO concentrations. A limitation with the bonding difficulty and interfacial interaction between the GO dispersed within the epoxy and the CF textile is blamed for the decrease in tensile and flexural strength. This is an ongoing study and these early stage results eludes to the potential to improve the epoxy/carbon fibre composites both tensile and flexural strength while still opening material tailoring opportunities for toughness and ductility improvements. Further investigation into the fabrication method is required to improve bonding and interfacial interaction between the GO dispersed within the epoxy and the carbon fibre textile.

5. Acknowledgments

The work is part of the DST-UKIERI collaboration project between Robert Gordon University and North Maharashtra University named Structural Reinforcement and Improvement of Compressive Strength after Impact (CIA) in Hybrid Graphene Nanocomposites. Kristof is also thankful for partial funding by the School of Engineering at Robert Gordon University for his studentship.

6. References

[1] Njuguna J, Pielichowski K, Desai S. Nanofiller-reinforced polymer nanocomposites. *Polym Adv Technol* 2008;**19**(8):947-959.

[2] Njuguna J, Pielichowski K, Alcock JR. Epoxy-Based Fibre Reinforced Nanocomposites. *Advanced Engineering Materials*. 2007 Oct 1;**9**(10):835-47.

[3] Nikfar B, Njuguna J. Compression-after-impact (CAI) performance of epoxycarbon fibre-reinforced nanocomposites using nanosilica and rubber particle enhancement. In IOP Conference Series: *Materials Science and Engineering* 2014 (Vol. **64**, No. 1, p. 012009). IOP Publishing.

[4] Khobragade PS, Hansora DP, Naik JB, Njuguna J, Mishra S. Preparation and analysis of multi-layered hybrid nanostructures. *Applied Clay Science*. 2016 Nov 30;**132**:668-74.

[5] Yang H, Shan C, Li F, Zhang Q, Han D, Niu L. Convenient preparation of tunably loaded chemically converted graphene oxide/epoxy resin nanocomposites from graphene oxide sheets through two-phase extraction. *Journal of Materials Chemistry* 2009;**19**(46):8856-8860.

[6] McAllister MJ, Li J, Adamson DH, Schniepp HC, Abdala AA, Liu J, et al. Single sheet functionalized graphene by oxidation and thermal expansion of graphite. *Chemistry of materials* 2007;**19**(18):4396-4404.

[7] Dreyer DR, Park S, Bielawski CW, Ruoff RS. The chemistry of graphene oxide. *Chem Soc Rev* 2010;**39**(1):228-240.

[8] Park S, Lee K, Bozoklu G, Cai W, Nguyen ST, Ruoff RS. Graphene oxide papers modified by divalent ions—enhancing mechanical properties via chemical cross-linking. *ACS nano* 2008;**2**(3):572-578.

[9] He H, Klinowski J, Forster M, Lerf A. A new structural model for graphite oxide. *Chemical physics letters* 1998;**287**(1):53-56.

[10] Desai S, Njuguna J. Graphite-based nanocomposites to enhance mechanical properties. In *Structural Nanocomposites* 2013 (pp. 61-74). Springer Berlin Heidelberg.

[11] Shen X, Pei X, Fu S, Friedrich K. Significantly modified tribological performance of epoxy nanocomposites at very low graphene oxide content. *Polymer* 2013;**54**(3):1234-1242.

[12] Wan Y, Tang L, Gong L, Yan D, Li Y, Wu L, et al. Grafting of epoxy chains onto graphene oxide for epoxy composites with improved mechanical and thermal properties. *Carbon* 2014;**69**:467-480.

[13] Bortz DR, Heras EG, Martin-Gullon I. Impressive fatigue life and fracture toughness improvements in graphene oxide/epoxy composites. *Macromolecules* 2011;**45**(1):238-245.

[14] Starost K, Njuguna J. A review on the effect of mechanical drilling on polymer nanocomposites. In IOP Conference Series: *Materials Science and Engineering* 2014 (Vol. **64**, No. 1, p. 012031). IOP Publishing.

[15] Hadden CM, Klimek-McDonald DR, Pineda EJ, King JA, Reichanadter AM, Miskioglu I, et al. Mechanical properties of graphene nanoplatelet/carbon fiber/epoxy hybrid composites: Multiscale modeling and experiments. *Carbon* 2015;**95**:100-112.

[16] Jiang S, Li Q, Wang J, He Z, Zhao Y, Kang M. Multiscale graphene oxide–carbon fiber reinforcements for advanced polyurethane composites. *Composites Part A: Applied Science and Manufacturing* 2016;**87**:1-9.

[17] Qin W, Vautard F, Drzal LT, Yu J. Mechanical and electrical properties of carbon fiber composites with incorporation of graphene nanoplatelets at the fiber–matrix interphase. *Composites Part B: Engineering* 2015;**69**:335-341.

[18] ISO B. 527-2: 1996. Plastics—determination of tensile properties—part 2: test conditions for moulding and extrusion plastics. British Standards Institution 1996:1-14.

[19] ISO B. 178: 2010. Plastics–Determination of flexural properties. International Standard Organization, Switzerland.

IOP Conf. Series: Materials Science and Engineering **195** (2017) 012010 doi:10.1088/1757-899X/195/1/012010

Designing of Hybrid Structured Glass Laminated Transparent Nano Composites through Vacuum infusion Technique

A.Mukherji*, N.D.Tarapure and G.N.Wakure

SP Advanced Engineering Materials Ltd, Mumbai, India

E-mail: drmukherji@gmail.com

Abstract. Glass is the most commonly used transparent material. However, glass is not suitable in applications where low weight, high strength is required. The present invention comprises a method of making a Transparent Glass Laminated Nano composite product. The product contains a Bi-directionally oriented E-Glass Fabric an essentially bidirectional yarn woven fabrics is stretched Bi-directionally by specially fabricated steel frame associated with both co and counter rotating device. These fibers include glass fibrics/cloths or mixtures of any of these. The synthetic fiber may be any synthetic silica based oven waived bi-directional or Uni-directional fabrics. Engaged gear provided in the device develops uniform tension on fabric, in both direction. Nano particle dispersed resin to be used is formulated with their respective curing agents and extenders. The formulated resin contains 0.1-0.5% of Nano additives and the product composed from 5-10 % of Glass fabric, between 10 to 20 % of ordinary glass, and between 60 - 80 % of the product is the Nano particles dispersed formulated resin, all measured by volume.

1. Introduction

Composite structures are increasingly being used in a wide variety of applications due to their high specific strength and modulus. Even though uses of composite structures can range from such fields as marine and automotive to skis and tennis rackets, their great strength-to-weight ratio makes composites ideal for aerospace applications. Glass has been commonly used for windows in airplanes because of its hardness, chemical inertness, abrasion resistance and relatively low costs. However, glass is brittle, heavy, shatters upon impact and difficult to form complex shapes. Light weight and mechanically strong alternatives are highly desirable. Optically transparent composite is a viable solution. Transparent composites can replace traditional heavy glass on the airplanes with light weight windows and windshields. Glass laminates can be used as windows and glaz¬ings in architectural and transportation applications, includ¬ing automobiles, rolling stock and airplanes. As used herein, a glazing is a transparent or semi-transparent part of a wall or other structure. Common types of glazings that are used in architectural and automotive applications include clear and tinted glass, including laminated glass. Laminated glazings comprising opposing glass sheets separated by a plasticized poly(vinyl butyral) (PVB) sheet, for example, can be used as windows, windshields, or sunroofs. In certain applications, glass laminates having high mechanical strength and sound-attenuating properties are desirable in order to provide a safe barrier with reducing sound transmission from external sources. A key function of an interlayer within a laminated glass structure is to prevent or at least minimise dispersion of dangerous glass fragments should the glass break.

Films made from plasticised PVB are considered to be particularly suited for use as an interlayer in laminated glass products due to their ability to strongly adhere with glass, their transparency and their sound physical and mechanical properties over a broad range of temperatures.
However, there are several following disadvantages to using plasticised PVB films as an interlayer:

- Films are relatively expensive.
- PVB films are also hygroscopic and must be handled under controlled atmospheric conditions to prevent excessive moisture uptake.
- Plasticised PVB film is often subject to blocking problems at the time of windup after film formation due to its very tacky surface. To prevent blocking, plasticised PVB films are typically provided with some form of parting means when stored or transported in the form of stacks of die-cut blanks or in the form of rolls. The use of PVB as an interlayer material therefore adds considerable expense to the final laminated glass product.
- PVB polymer itself is not particularly well suited to pigmentation or decoration. Polyvinylchloride (PVC) has been considered as a potential replacement for PVB as an interlayer material due to its following advantages:
- PVC is a commercial polymer that is readily available and relatively inexpensive.
- PVC can be easily calendared to a range of thicknesses, it can be plasticised to give a range of hardness values, it has suitable physical and mechanical properties for use as an interlayer and it can be pigmented or decorated to give a wide range of colours.
- PVC is not water sensitive and unlike PVB, does not require storage in a special controlled humidity environment.

However, the use of PVC in laminated glass structures has been limited due to the polymer having practically no inherent ability to adhere to glass. Attempts made to promote adhesion between glass and PVC within laminated glass structures
Thus, a transparent composite in which general-purpose glass filler is used and which shows high light transmissivity in a wide wavelength range is desired.
In view of vehicular applications, fuel economy is a function of vehicle weight. It is desirable, therefore, to reduce the weight of glazings for such applications without compromising strength and sound-attenuating properties. Desirable properties of such glazing should have sufficient strength to hold broken glass; sufficient adhesion to the glass to prevent or at least minimise dispersion of the broken glass; acceptable weather ability, etc. Normally thin interlayer polymers glass lamination process adapted so far. Normally production process involves two steps. In the first step the interlayer is placed between two sheets of glass and the resulting sandwich type structure is heated as it passes through nip rollers that squeeze the glass sheets into contact with the softened interlayer. A further step involves the treatment of the laminate so formed in an autoclave to complete the bonding process. US 4,121,014 discloses laminated windows containing adjacent layers of polycarbonate and plasticized polyvinyl acetal using adhesive film as polyurethane or an epoxy resin. But in case of polycarbonate, it's not laminated to glass. Also high temperature and high pressure is required for lamination in such case.
A transparent composite is based on the concept of matching the refractive index of the glass fibre with that of resin. Refractive index of the matrix can be modified to match with that of the glass fibre by changing the degree of polymerization of the polymer matrix. Various processing factors such as the amount of initiator, cross-linker, curing temperature and time must be carefully controlled to obtain a given refractive index. Significant reduction in weight can only be brought in by increasing the volume portion of polymer composite and reducing glass volume portion. But in doing so, one may not have desire strength to replace toughen Glass. Hence to get over from above limitation, moulded laminated reinforced composite glass by using glass fabric and nano composite may be the only solution. In view of transparency, refractive index plays a major role in polymer composites composed of more than two phases. Hence, polymer matrix with its given refractive index has to match the refractive index of the glass fibre. The refractive index of the glass fibre depends on its chemical composition, thermal history and fibre diameter. The diameter of the glass fibres pulled from a crucible of constant geometry depends upon the glass flow rate through the nozzle and the velocity at which the fibres are drawn by a take up wheel. Annealing can also increase the refractive index of chilled glass and could be potentially useful

in fabricating transparent composites. The thickness, TEX of fibre involved and the free voids appeared while fibres are wet in liquid polymer. A proprietary thermoset epoxy and polyester resin has been developed from Atul Pvt. Ltd and Satyen Polymers Pvt. Ltd. The resin has a low viscosity which allows it to be utilized in various liquid composite manufacturing processes. Several concepts of transparent composite were reported in literature. Olson, J. R., Day, D. E., Stoffer, J. O., 1992, "Fabrication and Mechanical Properties of an Optically Transparent Glass Fibre/Polymer Matrix Composite," Journal of Composite Materials, (Vol. 26, No. 8, pp. 1181–92) proposed transparent composite with glass reinforcements in thermoplastic Poly(methyl methacrylate) PMMA matrix. Kagawa, Y., Iba, H., Tanaka, M., Sato, H., and Chang, T., 1998, "Fabrication and Optical/Thermal Properties of Glass Particle-Epoxy Optically Transparent Composites," Acta Materialia, (Vol. 46, No. 1, pp. 265-271) manufactured mixing glass powder with the epoxy resin. Specialty rectangular glass ribbons were used as reinforcements by Chandrashekhara, K., Schuman, T., Sundararaman, S., Day, D., Velez, M., and Phan, N., 2008, "Manufacturing and Performance Evaluation of Glass-Ribbon Reinforced Transparent Composites," SAMPE International Symposium, (53, pp. 1-12) studied multiple density glass fibres in a polymeric matrix.

2. Materials

A clear polyester based one part resin system with refractive index matching the glass fibers has been developed at the SP advanced Materials Pvt. ltd. Ordinary Floating 1.5 mm thick glass ware used This resin system has been tailored to have a low viscosity thus enabling it to be used in a wide range of composite manufacturing processes. Bi-directional (0°/90°) woven glass fabric from Owens Corning Composite Materials, LLC, OH were used as fiber reinforcements. Epoxy from Atul Pvt. Ltd was used to coat the surface of the Glass.

3. Manufacturing

Manufacturing of the transparent composite was accomplished by the use of Under Gravity resin transfer and vacuum assisted resin transfer process. First method is engaged of a mould of desire size, made by thin and thick glass using phenol formaldehyde adhesive backing polyethylene cross-linked foam separated. In both the cases, two type of resins; formulated epoxy and Polyester are used. Respective curatives along with extenders used before transfer in the Mould. All leakage is sealed by liquid silicon glue. Resin transfer is done under gravity through flat hopper and a accumulator attached. The schematic of a typical first process as shown in Fig. 1. Wherein second method is engaged of tool comprises of two steel plates separated four separate steel spacers respectively. The assembly is laid on the station of the isolated steel chamber. Resin is charged thorough vacuum assisted technique. The schematic of a typical VARTM process is shown in Fig. 2. The infusion of resin into the fabric can be influenced by the permeability of the preform, the fiber architecture, and the presence of fabric crimping. The first step in the manufacturing process is to prepare the mold. The mold is thoroughly sanded to remove any scratches and cleaned using solvents and cotton cloths. The glass fabrics are stretched using tool which is then laid on the mold. Glass fabric is used in only vacuum assisted resin transfer method.

Entire assembly was placed in a vacuum chamber which was sealed around its perimeter with a general use tacky tape. The resin inlet and vacuum lines were positioned at opposite end of the vacuum chamber. The vacuum line is connected to a vacuum pump and the mold is checked for leaks. Any leaks which are found must be sealed. Once the mold preparation is complete, the resin is degassed by casting it on the hopper of Flat base. In the Resin formulation, Menthol was added to reduce viscosity. Volatile methanol will escape out leaving behind bubble free resin. The bubble and air free resin was accumulated and discharged into the mould assemble. The degassed resin is now allowed to flow through the inlet line. Added extender maintains the desire viscosity for flow and gets cure over a period in presence of initiator. The resin transfer process can take approximately ten to fifteen minutes to complete. Once the part is fully filled and gelation starts vacuum lines was opened.

Both the inlet and vacuum lines completely shut-off throughout the gelling. This keeps the resin within the part and prevents voids from forming within the part which in turn will affect the transparency of the finished composite. During curing chilled air passed to reduce the exothermic reaction heat. Once cured, the part were tpost cured. The process involves heating the mold to 240 °F for 2 hours. The cure cycle used for the part manufacturing shown in Fig. 2. After the completion of post cure cycle, the mold cooled to room temperature and the part separated from the mold. The part were then be examined for voids, surface finish, and the refractive indices matched.

4. Results & Discussion

Method -1

In this example a flat Glass laminated Transparent Nano composites product made having a cross section 8 x 1000 of length 1000 mm. This product was normally made with 35% Ordinary Glass in two layers, and 65% resin, all measured by volume. Two type of Ordinary and Toughen Glasses of different thickness, and two types of resin with varying chemical formulation used in making the products of the example. Glass inside surfaces treated chemically with different coupling agents.

Table-1:Composites properties

Types of Fabrics	Degree of Inclination		Heat of Reaction	Transmittance in visible light I/I0 (a.u)	Surface indolence	Air Bubble
Ordinary 1.5mm Glass (without surface treatment)	(A1)	50	160	92	Low	Low
	(A2)	100	165	90	Moderate	Moderate
	(A3)	150	170	86	Higher Side	Higher
Ordinary 3mm (AIES) Treated Glass	(B1)	50	166	89	Low	Low
	(B2)	100	167	87	Moderate	Moderate
	(B3)	150	170	85	Higher Side	Higher
Ordinary 6 mm (TDI) treated Glass	(C1)	50	162	89	Low	Low
	(C2)	100	165	87	Moderate	Moderate
	(C3)	150	170	85	Higher Side	Higher
Toughened 6 mm (AIES) Treated Glass	(D1)	50	166	89	Low	Low
	(D2)	100	165	87	Moderate	Moderate
	(D3)	150	168	85	Higher Side	Higher

IOP Conf. Series: Materials Science and Engineering **195** (2017) 012010 doi:10.1088/1757-899X/195/1/012010

Toughened and non-toughened less expensive glasses of thickness range from 1.5-6 mm are treated with various coupling agents like Amino isopropyl ethoxy silane, isopropyl triisostearoyl titanate (abbreviated as titanate), g - aminopropyl trimethoxy silane (abbreviated as silane), sebacoyl chloride (SC), and toluene diisocynate (TDI). Treatment done at ambient temperature. More typically, Amino isopropyl ethoxy silane and toluene diisocynate (TDI) surface treating agents are used in the present example, to improve Glass to resin bonding.

Under Gravity resin is charged from resin chamber to the pre-fabricated mould through a nozzle. The product made is resin transfer without pre-tensioned Glass fabrics. The charging of resin executed by maintaining different degree of inclination. The degree of inclination ranges from 5 to 20o. Degree of Inclination has direct impact on thickness Variation and curing Mechanisms of the resin. More inclined charging of resin results in bulging of resin at the center part of the Mould and uncontrolled process due to leakage problem. Higher inclination leads to faster curing and high exothermic reaction. Degree of inclination and surface treatments imparts effects on various properties of the Glass as showed in Table1. The products made are represented in Fig1 where Transparent Nano composites indicated at C1,C2,C3 and D1,D2,D3 were made respectively with treated and Untreated Glass at different degree of inclination. Mould positioned approximately as shown in the drawing.

A first set of trials made with the three types of surface treatment Glass used for isolated pre-fabricated Mould wherein polyester resin was discharged manually. A second set of trials were made using a clear cast epoxy resin. Polyester resin employed was developed by Satyen Polymes. Pvt. Ltd. and Epoxy by Atul Chemicals Pvt. Ltd.. Nano particles like carbon tubes/nano clay/nano silica dispersed and exfoliated in the base resins prior to specific formulation. Base resins formulated of minimum shrinkage and higher toughness. The scope of invention includes the use of extender like vinyl Acetate, Epoxy and Methanol in the formulation to reduce the heat of reaction. Type and dose of Nano materials and extenders in formulation plays important role on transparency, heat of reaction and Bonding of Glass to Resin results as showed in Table 2 .

The products made with Nano material and with both resins were excellent in appearance. The discharging clear resin by manual technique has tendency of Bubble formation and bulging effect at the center on higher inclination.

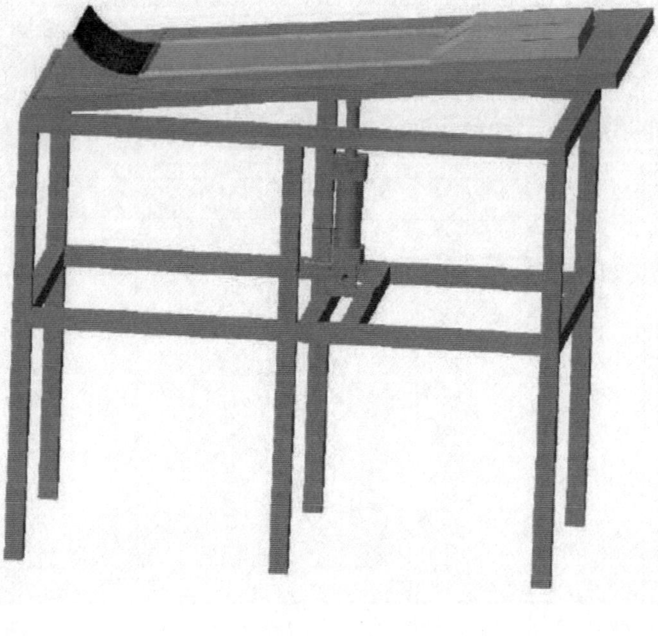

Fig.1 Schematic 3D image of hydraulically actuated Station Used to charge resin

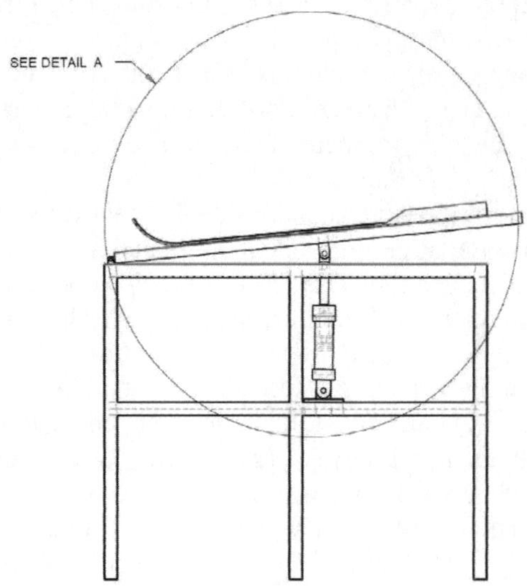

Fig.2 Schematic 2D image of hydraulically actuated Station Used to charge resin

Method - 2

Synthetic fabrics like polyester, Basalt or Glass or more typically E-Glass fabrics having a 80-100 g/m^2 basis weight more typically 90g/m^2, a density of about 70 kg/m^3 and a thickness between 0.5mm and 1 mm was used. Very thin coating is made on glass by various chemicals like Epoxy, Polyurethane, SBR (Styrene butadiene Rubber) latex; NBR (Nitrile butadiene Rubber), NR (natural rubber), Polybutadine, Solvent based polyvinyl acetate, Polymethyl methacrylate, Polyethyle acrylate, styrene and acrylonitrile and acrylamide in the presence of different redox initiator systems such as vanadium - cyclohexanol, vanadium - cyclohexanone, etc. More typically, Epoxy, Polyurethane and Polyvinyl acetate surface coating used in the present example, to improve Glass to resin bonding. Thick and flexible coating on Glass surface is more vulnerable in holding cure resin on breakage of the product wherein rigid and thin coating holds resin profoundly as showed in Table 2 and 3. Wherein thin flexible coating on top of treated glass surface holds thin Glass broken fragment Glass but has significantly reduce transparency and Mechanical properties.

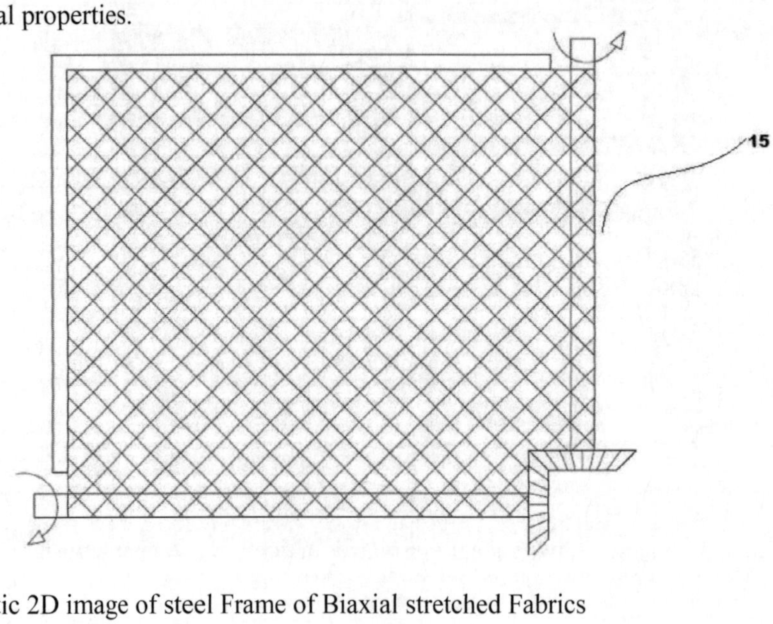

Fig.3 Schematic 2D image of steel Frame of Biaxial stretched Fabrics

IOP Conf. Series: Materials Science and Engineering **195** (2017) 012010 doi:10.1088/1757-899X/195/1/012010

Table: 2 Bond features and heat reaction

Formulation: Polyester-100 Phr, cobalt octoate -2%, Methyl Ethyl Ketone peroxide (MEKP)-1.5 %	Weight % used						Heat of Reaction	Bonding Between Glass and Resin
	Nano Materials			Extender			Temperature (0C)	
	Carbon Nano Tube		Clay Nano	vinyl acetate	Methanol	Epoxy		
Ordinary Glass without any coating and treatment	(E)	0.1%	0.2%	2%	0 %	4%	160	Broken into Fragments
Polyurethane Coated ordinary Glass	(F1)	0%	0.5%	2%	0.0%	5.0%	160	Broken into Fragments
	(F2)	0.3%	0%	4%	2%	0.0%	150	Broken into Fragments
Polyvinyl Acetate Coated Toughened Glass	(G1)	0.2%	0%	2%	1.0%	5.0%	150	Broken into Fragments but does not come out
	(G2)	0.0 %	0.7%	4%	2%	0.0%	160	Broken into Fragments but does not come out
Methyl acrylate Coated Toughened Glass	(H1)	0.2%	0.0%	2%	2%	5.0%	145	Broken into Fragments
	(H2)	0.0%	0.7%	4%	2%	0.0%	165	Broken into Fragments

IOP Conf. Series: Materials Science and Engineering **195** (2017) 012010 doi:10.1088/1757-899X/195/1/012010

Table-3:Bonding and light transmittance properties

Formulation: Epoxy-50% Hardner-50%	Weight % used					Transmittance in visible light I/I0 (a.u)	Bonding Between Glass and Resin
	Nano Materials			Extender			
	Carbon Nano Tube		Clay Nano	Polyurethane	Polyvinyl ester		
Polyurethane Coated ordinary Glass	(I1)	0.2%	0%	2%	1.0%	88	Broken into Fragments
	(I2)	0.0 %	0.7%	4%	2%	86	Broken into Fragments
Polyvinyl acetate Coated Toughened Glass	(J1)	0.2%	0%	2%	1.0%	86	Broken into Fragments
	(J2)	0.0 %	0.7%	4%	2%	82	Broken into Fragments
Epoxy Coated Toughened Glass	(K1)	0.2%	0%	2%	1.0%	85	Broken into Fragments
	(K2)	0.0 %	0.7%	4%	2%	88	Broken into Fragments

First set of trial made with two types of fabrics and one UD roving pre-stretched in the steel frame placed in the specific mould wherein polyester resin discharged under vacuum assisted Method. Second set made with formulated clear cast epoxy resin. Mould was kept in a vacuum pump facilitated isolated chamber where in resin discharge process was developed. The fabrics developed and supplied by Woon's Corning. Scope of invention includes Synthetic fibers like Basalt fabrics, Glass fabrics, Aramide, polyproylene and polyester Fabrics. More typically, Basalt fabrics, Glass fabrics and polyester fabrics are used in the present example. Fabrics GSM ranges from 70-400g/m^2 and more typically 80-200g/m2. Higher GSM fabrics results in a significant reduction in transparency but enhancement in resistance to compressive load, higher stiffness and bending strength as well.

Pre-stretched fabrics contributes major role in enhancing compressive load bearing property of the product wherein fabrics GSM and text of fibers involved controls transparency of the product as given in Table-4.

The vacuum chamber of 1000-2400mm length, 1200mm width and 800mm of depth used to produce Glass laminated Transparent Nano Composites of 10-20 mm thick, 1000 mm width and of 1500 mm length. The Chember is Vacuumed to a level of 0.12 mm Mercury. The trials were run after achieving a vacuum level and positioning the Mould assemble at 0 o to Base. Hydraulic actuation of the base station where Mould assemble is placed, started clamping as against top of the vacuum chamber before opening valve to discharge resin. The mould assemble includes the steel spacer separating two halves of the mould. Two sets of independent steel spacer attached separately to respective Mould halves holds the pre-tensioned fabrics. Pre-tensioned fabrics placed at the center of the Mould assemble separates Mould halves equally. Removing air from the isolated chamber through Vacuum Pump attached. This process significantly reduced air bubbles, thickness variation and has controlled cure mechanism. The resin was standard clear cast grade polyester obtained from Satyen Polymers. The resin is charged from resin

IOP Conf. Series: Materials Science and Engineering **195** (2017) 012010 doi:10.1088/1757-899X/195/1/012010

chamber through a valve attached nozzle. The product made by Vacuum assisted resin transfer with and without pre-tensioned Glass fabrics. Vacuum level is maintained between 0.8-0.3 mm as are shown in the Table 5. In the present example Glass fabrics between 80-200 GSM are also replaced with 70 - 200 GSM polyester and Basalt fabrics.

The products made are represented in Fig. 3, 4 and 5 where transparent Nano composites indicated at A1, A2, A3 and B1, B2, B3 were made respectively with Glass fabrics, Basalt fabric and Polyester Fabrics with polyester and Epoxy as cast resin. The thin glass surface is also treated with different surface treating agents to have bonding with rein as indicated in Table 4. Suface treatment of Glass surface and use of Glass fabrics have significant impact on Transparency and mechanical Property of the Product.

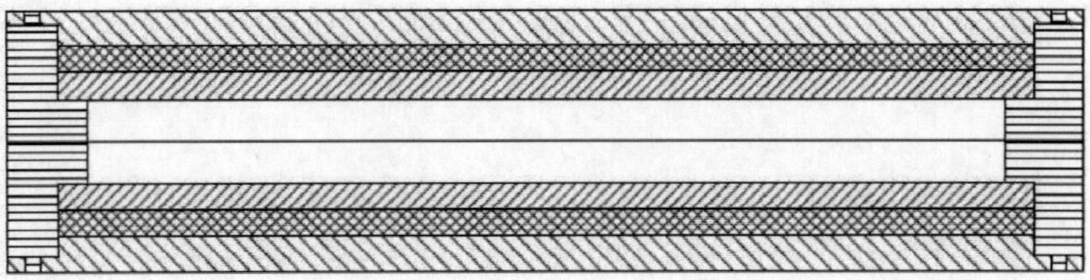

Fig.4 Schematic Sectional View of Mould used in Glass composites manufacturing

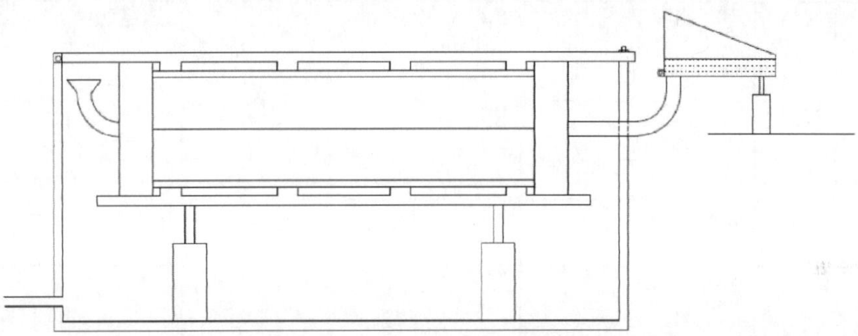

Fig.5- Schematic 2D image of Hydraulically actuated Station Used to charge resin Under Vaccum Infusion Technique.

The present invention relates to an improved glass laminated nano-composite which is very durable, higher mechanical properties compared to conventional glass and heat insulative yet very light weight. The present invention also provides a method of producing the same in very simple and economic manner by using very less expensive glass as a starting material and various easily available resins.

The basic resins are used for this purpose are suitable mixtures of unsaturated polyester, Vinyl ester and clear cast Epoxy and their blends. The resins formulated before using to minimize shrinkage and higher toughened properties.

IOP Conf. Series: Materials Science and Engineering **195** (2017) 012010 doi:10.1088/1757-899X/195/1/012010

Table 4 Impact performance

Epoxy coated 1.5 mm Ordinary Glass at Vacuume 200							
Types of Fabrics	GSM		Bending Strength N/mm2	Transmittance in visible light I/I0 (a.u)	Surface indolence +/- mm	Ballistic 80 m/s speed 3J	Air Bubble
E-Glass Fabrics	(L1)	80	70	90		1.2 au	95% Micro voids free
	(L2)	120	73	86	+/-2	1.03 au	90% Micro voids free
	(L3)	200	80	84	+/-2	1.04 au	87 Micro voids free
E-Glass UD (A)	(M1)	200	40	82	+/-2	2.5 au	80 Micro voids free
Basalt Fabrics	(N1)	80	70	88	+/-2	1.02 au	90% Micro voids free
	(N2)	120	78	85	+/-2	1.5 au	94% Micro voids free
	(N3)	200	87	83	+/-2	1.3 au	90% Micro voids free
Polyester	(O1)	180	40	72	+/-2	2.5 au	60% Micro voids free
	(O2)	200	40	74	+/-2	3.5 au	65% Micro voids free

Table-5: Mechanical properties

Types of Fabrics	Vacuum Level		Compressive Strength	Transmittance in visible light I/I0 (a.u)	Surface indolence +/- mm	Drop Impact (50GM) At 20 ft Height level at 10 sites	Air Bubble
E-Glass Fabrics of 80 GSM under pre-tensioned condition	(P1)	80	70	89	+/-1.05	Damage quantification (2 au)	Blister free
	(P2)	120	73	90	+/-1.02	(1.5au)	Blister free
	(P3)	200	80	92	+/-0.5	(1 au)	Blister free

5. Conclusion

The cured moulded laminated reinforced composite glasswere tested for density, Glass to resin bond strength, tensile, drop impact and flexural properties. These tests showed that the composites glass possess good physical and mechanical properties. These new composite glass has applications in windshields, windows and other components where a strong, lightweight transparent material is desirable. These moulded laminated reinforced composite glass will also have potential applications as backing material in transparent armor systems. These cured composite glass are low density product as compared to normal Glass. Also, the laminated composite glass of the invention has improved glass to resin bond strength which helps hold the glass fragments on breakage. The laminated composite glass has excellent weather ability or environment resistance and has excellent laceration property as compared to normal composites glass. Thus the present invention provided an moulded laminated reinforced composite glass which is very durable, higher mechanical properties compared to conventional glass and heat insulative yet very light in weight. The present invention also provides a process of producing the same in very simple and economic manner by using very less expensive glass as a starting material and various easily available resins.

6. References

[1] Olson, J. R., Day, D. E., Stoffer, J. O., 1992, "Fabrication and Mechanical Properties of an Optically Transparent Glass Fiber/Polymer Matrix Composite," *Journal of Composite Materials*, Vol. **26**, No. 8, pp. 1181–92.

[2] Kagawa, Y., Iba, H., Tanaka, M., Sato, H., and Chang, T., 1998, "Fabrication and Optical/Thermal Properties of Glass Particle-Epoxy Optically Transparent Composites," *Acta Materialia*, Vol. **46**, No. 1, pp. 265-271.

[3] Agarwal, B. D., Broutman, L. J., and Chandrashekhara, K., 2006, "Analysis and Performance of Fiber Composites," John Wiley & Sons, Inc., Hoboken, NJ, Third Edition.

[4] Menta, V. G. K., Sundararaman, S., Chandrashekhara, K., Phan, N., and Nguyen, T., 2008, "Hybrid Composites using Out-of-Autoclave Process for Aerospace Sub-structures," SAMPE International Symposium, **53**, pp. 1-11.

[5] Iba, H., Chang, T., Kagawa, Y., 2002, "Optically Transparent Continuous Glass Fiber-reinforced Epoxy Matrix Composite: Fabrication, Optical and Mechanical Properties," *Composites Science and Technology*, Vol. **62**, pp. 2043-2052.

[6] Patel, P.J., Dehmer, P.G., and McCauley, J. W., 2000, "Transparent Armor," *Advanced Materials and Process Technology*, Vol. **4**, No. 3, pp. 1-5.

[7] Chandrashekhara, K., Schuman, T., Sundararaman, S., Day, D., Velez, M., and Phan, N., 2008, "Manufacturing and Performance Evaluation of Glass-Ribbon Reinforced Transparent Composites," SAMPE International Symposium, **53**, pp. 1-12.

[8] Wilenski, M. S., Markus, A. M., Godby, L. A., 2009, "Varied Glass Density Reinforcement of Composites," U.S. Patent 057494.

IOP Conf. Series: Materials Science and Engineering **195** (2017) 012011 doi:10.1088/1757-899X/195/1/012011

Environmental Particle Emissions due to Automated Drilling of Polypropylene Composites and Nanocomposites Reinforced with Talc, Montmorillonite and Wollastonite

K Starost[1*], E Frijns[2], J V Laer[2], N Faisal[1], A Egizabal[3], C Elizextea[3], I Nelissen[2], M Blazquez[4] and J Njuguna[1*]

[1] Centre for Advanced Engineering Materials, School of Engineering, Robert Gordon University, Aberdeen, AB10 7GJ, United Kingdom
[2] VITO NV, Boeratang 200, 2400 Mol, Belgium
[3] TECNALIA, Mikeletegi pasealekua, 2 E-20009, Donostia Spain
[4] INKOA SISTEMAS, SL. Ribera de Axpe 11, Edificio D1, Dpto 208. 48950, Erandio, Bizkaia Spain

E-mail: k.starost@rgu.ac.uk, j.njuguna@rgu.ac.uk

Abstract. In this study, the effect on nanoparticle emissions due to drilling on Polypropylene (PP) reinforced with 20% talc, 5% montmorillonite (MMT) and 5% Wollastonite (WO) is investigated. The study is the first to explore the nanoparticle release from WO and talc reinforced composites and compares the results to previously researched MMT. With 5% WO, equivalent tensile properties with a 10 % weight reduction were obtained relative to the reference 20% talc sample. The materials were fabricated through injection moulding. The nanorelease studies were undertaken using the controlled drilling methodology for nanoparticle exposure assessment developed within the European Commission funded SIRENA Life 11 ENV/ES/506 project. Measurements were taken using CPC and DMS50 equipment for real-time characterization and measurements. The particle number concentration (of particles <1000nm) and particle size distribution (4.87nm – 562.34nm) of the particles emitted during drilling were evaluated to investigate the effect of the silicate fillers on the particles released. The nano-filled samples exhibited a 33% decrease (MMT sample) or a 30% increase (WO sample) on the average particle number concentration released in comparison to the neat polypropylene sample. The size distribution data displayed a substantial percentage of the particles released from the PP, PP/WO and PP/MMT samples to be between 5-20nm, whereas the PP/talc sample emitted larger particle diameters.

1. Introduction

The use of sillicate nanofillers as mechanical reinforcements in polymers is increasingly being well established throughout literature. This has generated an influx into various high performance lightweight-material commercial industries such as the automotive industry [1]. To continue to improve performance and economical costs, industries are using nano-fillers to reinforce the composite materials. talc [2], montmorillonite (MMT) [3] and Wollastonite (WO) [4] are commercially available fillers increasingly being researched and introduced in the automotive industry. These micro and nano-sized fillers have however also shown potential cytotoxicity if exposed and inhaled [5-7]. However, there is still an insufficient understanding on how these fillers effect the release of nanoparticles to evaluate and quantify the full risks associated to the emissions and nanoparticle exposure into the environment [8,9]. The results demonstrated within this paper are part of an ongoing study looking into the release and exposure of nanocomposite materials when under a simulated and controlled life cycle scenario: drilling process. This paper reveals some of the findings from PP-based nanocomposite materials reinforced with talc, MMT and WO. As to validate the improved properties and to support the link between mechanical performance and the nanoparticle release of the materials, the samples underwent a tensile test in accordance to ISO 527 [15] at a 2mm/min standard. The PP/WO sample demonstrated the same mechanical performance of the reference sample (PP/talc) with a 10% density decrease.

2. Experiments

2.1. Materials and Fabrication

A commercially available Polypropylene homopolymer was purchased from Lyondell Basell Industries (Moplen HP648T). A 20% talc filled polypropylene copolymer (Holstacom XM 2416) was also chosen from Lydonell Basell Industries as another reference material. The reinforcements and concentrations chosen were 5 wt. % Wollastonite (WO) from Nordkalk (Harwoll 7ST5) and 5 wt. % of montmorillonite (MMT) from Nanocor Corporation (Nanomer I30T).

The Coperion ZSK 26 MEGAcompounder twin-screw extruder was used for homogenization of the nanocomposites. The extruded pellets of the materials were moulded by injection process by means of an Arburg All Rounder 270C-300-100 Injection Machine. Due to the diverse polarity nature of the polypropylene and the MMT and WO, a coupling agent (POLYBOND 3200 from ADDIVANT) was used to ensure adhesion between the nanofillers and the polymer. Therefore, four sets of samples were fabricated: neat PP, PP with 20% talc, PP with 5wt. % MMT and 2 wt. % coupling agent, and PP with 5wt. % WO with 2 wt. % coupling agent. A common sample size of 70x45x5mm were prepared for the drilling investigations. The corresponding dog-bone standard sample was fabricated for the polymer reference standard ISO 527 tensile test [15].

2.2. Automated Drilling Procedure

The materials were tested using a purpose built controlled test chamber that allows direct measurement of nanoparticles emitted during drilling. The process is developed and initiated by the SIRENA Life project –an acronym for Simulation of the Release of Nanomaterials from Consumer Products for Environmental Exposure Assessment. This process is designed to simulate mechanical drilling on nanocomposite materials and is continued work from a previous European Commission funded NEPHH project titled 'Nanomaterials related Environmental Pollution and Health Hazards throughout their life cycle' (NEPHH, Project No. 228536) delivered by the same team of researchers [10]. The methodology allows for a categorical representation of the nanoparticles released from the material without any background interference and in higher accuracy than reported in the literature.

Based on industrial specifications and previous studies carried out on nanocomposite drilling, a standard Dremel 4000 drilling tool with an industrial standard stainless steel 3.5mm twist drill bit was used at 10000 rpm with a feed rate of 78mm/min [10-13]. The setup uses an automated drilling assembly operated externally to the chamber to permit a repeatable and controlled environment within the chamber as shown in Figure 1.

The closed steel chamber has dimensions of 740 mm x 550 mm x 590mm, and therefore a total inner volume of 0.240m^3. It is designed to assure a closed environment to simulate an appropriate volume around the drill and minimising electrostatic attraction to the surfaces. To quantify only the particles released from the sample, the chamber was initially cleared of particles through an inflow of clean air with the use of TSI 99.97% retention HEPA Capsule Filters. A separate capsule was constructed around the drill with separate air flow to avoid any interference of the drilling fumes on the particle number concentration within the capsule. The clean air system using the HEPA Capsule filters was capable of producing a particle number concentration reading within the chamber of 0 #/cm^3 as measured using a Condensation Particle Counter, CPC model 3783 at a flow rate of 0.6 litres per minute (lpm).

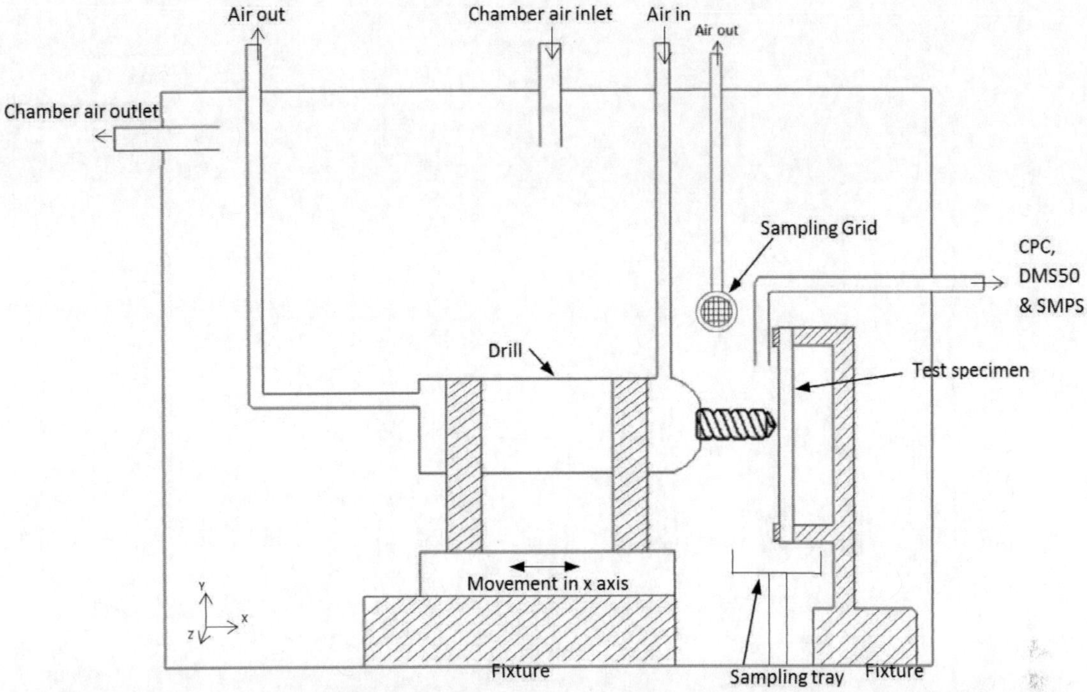

Figure 1: Schematic of automated drilling process setup within test chamber to remove background interfering particles and allow for a repeatable and controllable test procedure

An outlet channel is placed adjacent to the test specimen for the nanoparticle release equipment readings. A sampling grid for post-test analysis and characterization of the airborne particles was placed next the test specimen with a slight suction to attract and prevent particles from detaching away from the grid. An additional sampling tray was positioned below the test specimen for collection of the deposited particles for further post-test analysis.

A Cambustion DMS50 Fast Particle Size Spectrometer with a 1 second sampling period, inlet flow rate of 6lpm, with 34 distinct particle diameters of size range between 4.87nm – 562.34nm was used for the particle size distribution. This allowed for a size distribution every second compared to an SMPS of 55s and therefore an accurate representation of the particles being released from the sample in a given time. Particles released from the drilling were sampled as shown in Figure 1 to be analysed with an SEM, EDX and XRF were used but are beyond the scope of this paper.

3. Results and Discussion

3.1. Filler Effect on Particle Number Concentration

The polypropylene based nanocomposite samples underwent the automated drilling procedure described. Each test consisted of drilling eight holes within 3 minutes followed by 1 minute of no drilling. This methodology allows for both an investigation into the particles released at the instant of drilling and the remaining emissions airborne post drilling. Using the CPC, the particle number concentration was quantified in situ with a sampling rate of 1 second. An average of the repeated test for each sample is displayed in Figure 2.

The peaks observed in Figure 2 clearly exemplifies the eight holes drilled within the 3 minutes for the four PP based samples. On most of the peaks, the movement of the drill going in and out of the sample can also be seen from peaks being faintly divided into two peaks. When the drill is out of the sample, the particle number concentration is seen to drop between each hole being drilled. The particle number concentration can be perceived to then relatively stabilize during the 1 minute after the drilling has ended, but does not drop back to the initial 0 #/cm^3. Thus, the particles produced from the drilling remain airborne within the chamber environment.

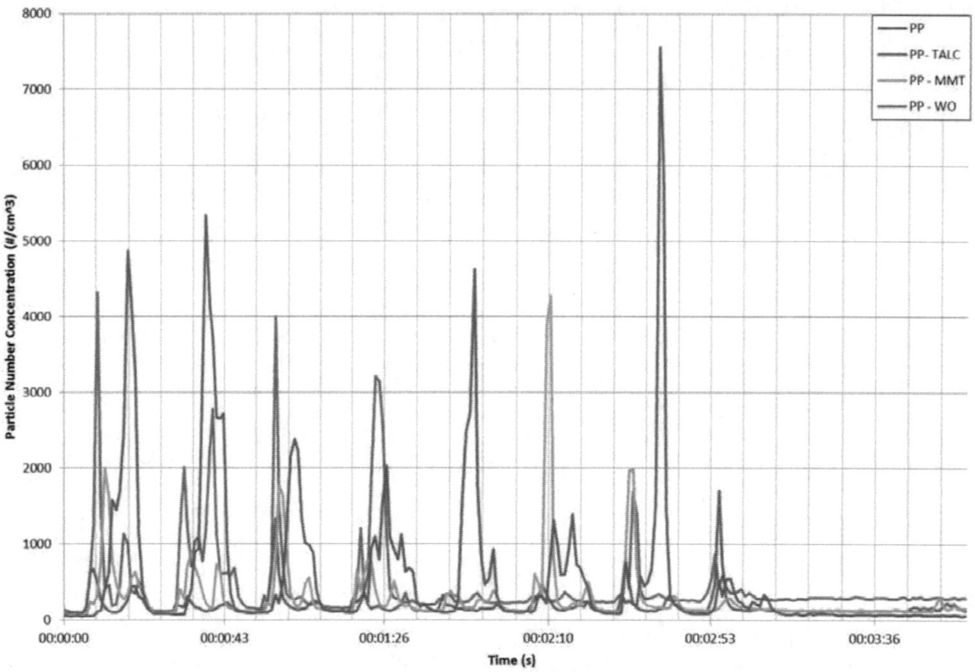

Figure 2: Particle number concentration of PP based nanocomposite samples during eight holes drilled within 3 minutes followed by 1 minute of no drilling

The PP/WO sample demonstrated the largest average peaks across the eight holes drilled when compared to the PP, PP/talc and PP/MMT. The PP and PP/MMT sample displayed similar size peaks during the drilling, whereas the PP/talc sample indicated the lowest peaks of all the samples. However, the PP/WO sample demonstrated to have the lowest particle number concentration at the end of the four-minute sampling period, and in contrast, the PP sample displayed the highest particle number concentration. Although, the PP/WO sample illustrated to have the highest peak value, peak average and total average over the entire four-minute sampling period, the sample presented the least particle number concentration at the end of the four minutes. The PP/WO released particles are therefore, perceived to deposit quicker than the three other samples. This conflicts with the nano-reinforced samples having a lower density to the PP or PP/talc sample. The lower particle number concentration after drilling is beneficial in relation to nanosafety and if considering materials safer by design, but the cause is ambiguous. The particles suggest being more reactive and either attracted to components within the chamber or agglomerating to larger particles the CPC is unable to pick up.
In relation to the average particle number concentration over the sampling period, PP/WO is the only sample that produced an increase in particles over the PP sample, with a 30% increase, compared to the decrease of 59% and 33% from the PP/talc and PP/MMT samples respectively. The nano-filled samples therefore, exhibited a converse 33% decrease (PP/MMT) or a 30% increase (PP/WO) on the particle number concentration released over the PP sample. However, these sets of results prominently indicate that the matrix has a substantial contributing factor on the particle number concentration when comparing the PP samples with other polymers. A similar trend with a silicate nanofiller producing the most particles during drilling and the influencing factor of the PP is observed in the NEPHH project reported in Irfan et al 2013 [14].

3.2. Filler Effect on Particle Size Distribution

Simultaneous to the data gathered for the particle number concentration, the particle size distribution was quantified in situ using a DMS50. This provides a better understanding of the size of the particles number concentration seen in the Figure 2. Additional to the DMS50, an SMPS was also utilised to evaluate the size distribution (not reported). In comparison to the SMPS which has a sampling period of

1 minute, the DMS50 generates a size distribution every second. This provides a more live visual of the nanoparticles as they are being released from the material before the particles are dispersed within the chamber.

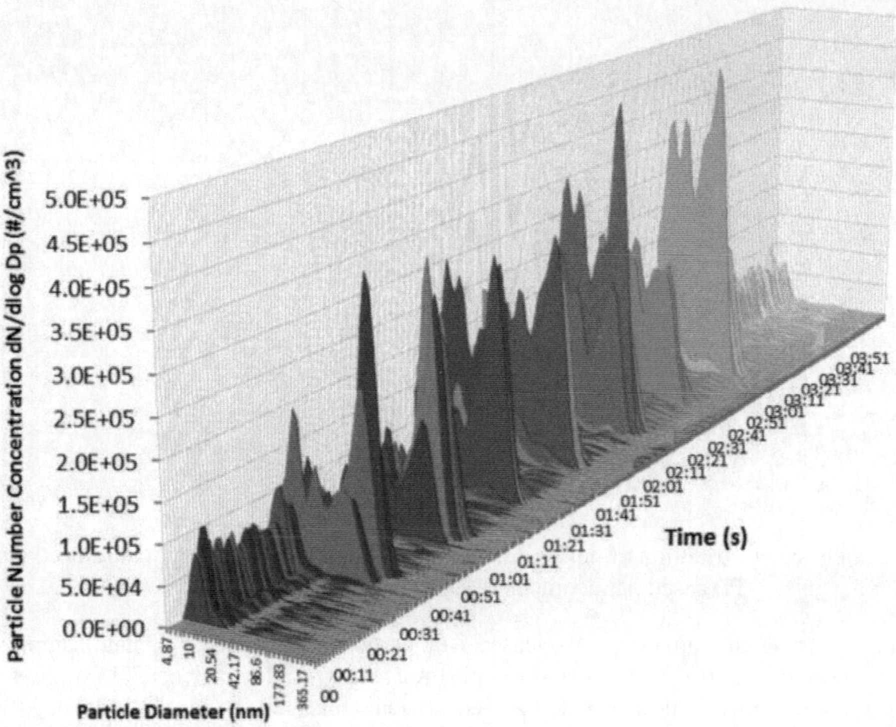

Figure 3: Particle size distribution over four minutes of PP/talc sample as measured on the DMS50

Since a size distribution is generated every second, Figure 3 illustrates the combination of the particle number concentration and its corresponding size distribution in a three-dimensional plot over the 4-minute sampling period. The drilling of the eight holes is perceivable with an initial introduction of particles for the first hole followed by 7 substantial peak concentration of particles emitted for the remaining holes. The size distributions between peaks and after drilling are less visible due to the high concentrations from the peaks. As indicated on the CPC data displayed in Figure 2, this highlights the vast particle concentrations produced at the time of drilling before the emissions disperse within the chamber and stabilise. Although the particles do stabilise and reduce in particle number concentration, a small percentage (<400#/cm^3) still remain airborne within the chamber environment. Figure 3 also demonstrates that the peaks of particles generated due to the drilling across the eight holes are relatively consistent in particle diameter. It is important to note that the data is taken from a separate run to the CPC and SMPS data due to the required increased inflow rate (6 lpm for the DMS50 compared to 0.6 lpm for the CPC) which is the probable cause for the increase in particle number concentrations relative to the CPC data represented in Figure 2.

Similar three-dimensional plots as illustrated in Figure 3 were generated for the four samples. In order to allow for a comprehensible comparison between the samples, a two-dimensional plot of the size distribution taken from the highest peak for each sample is displayed in Figure 4. The PP, PP/WO and PP/MMT samples revealed a substantial percentage of their particles between 5-20 nm particle diameter range. Therefore, the PP, PP/WO and PP/MMT appeared to release a greater proportion of particles with smaller diameters compared to the PP/talc sample. Despite exhibiting a peak at a greater particle diameter, it must be noted that the PP/talc sample released a high peak concentration of particles within the same diameters of other three PP-based samples. The data therefore suggests that the WO and MMT nano-sized reinforcements have little effect on the particle size distribution. The increase in particle number concentration seen in Figure 2 could be due to larger particle diameters as the CPC has a size range between 7-1000nm.

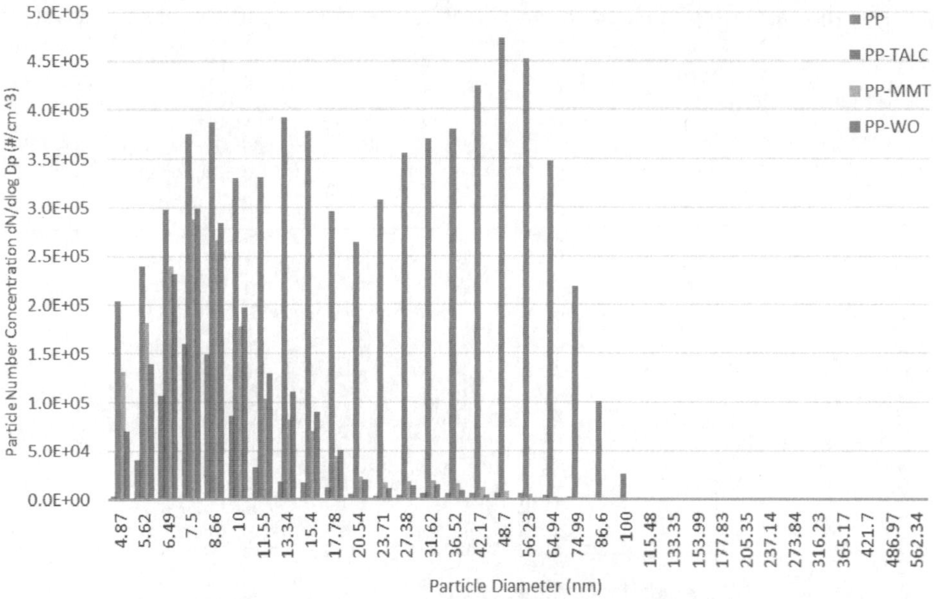

Figure 4: Particle size distribution of peak number concentrations during 4-minute sampling period for PP based nanocomposite samples recorded on DMS50

Figure 4 demonstrates that all samples released nanoparticles during the 4-minute sampling period, including the neat PP sample. None of the samples released particles between 115-562nm. The data from the particle size distribution reveals that the particles released are highly influenced by the PP matrix. The nano-reinforcements of WO and MMT did not demonstrate any additional nano-sized peaks in the DMS50 or the SMPS results, and must therefore be agglomerating or adhering to the matrix. The talcum reinforcement is the only filler showing an effect on the particle size distribution. This could also be due to the higher percentage of filler concentration. A further investigation is required to understand the nature of the larger particle diameter introduced by the PP/talc sample.

4. Conclusion

The automated drilling process validates a nanoparticle release testing methodology permitting a direct measurement of nanoparticle emissions into a clean chamber environment without any background interference. Talc and WO reinforced composites have both demonstrated nanoparticle release for the first time and compared to MMT. The initial data presented reveals minor difference in nanoparticle release between the four PP-based samples. All four samples exposed a concentration of nanoparticles introduced due to the drilling into the chamber environment. The nanofillers (WO and MMT) demonstrated both an increase and decrease in nanoparticle release, but no visible difference in particle size distribution. The higher concentration of talc as a filler had the biggest effect on particle size distribution compared to the other PP-based samples. The data presented is part of an ongoing study which will further investigate the initial findings and understand the causality of the results.

5. Acknowledgments

The work is funded by and part of the European Commission Life project named Simulation of the release of nanomaterials from consumer products for environmental exposure assessment (SIRENA, Pr. No. LIFE 11 ENV/ES/596). The access and use of the facilities at the Flemish Institute for Technological Research (VITO) was funded by QualityNano Project through Transnational Access (TA Application VITO-TAF-382 and VITO-TAF-500) under the European Commission, Grant Agreement No: INFRA-2010-262163. Kristof is also thankful for partial funding by the School of Engineering at Robert Gordon University for his studentship.

IOP Conf. Series: Materials Science and Engineering **195** (2017) 012011 doi:10.1088/1757-899X/195/1/012011

6. References

[1] Njuguna J, Pielichowski K, Desai S. Nanofiller-reinforced polymer nanocomposites. *Polym Adv Technol* 2008;**19**(8):947-959.

[2] Lapcik Jr L, Jindrova P, Lapcikova B. Effect of Talc Filler Content on Poly (Propylene) Composite Mechanical Properties. *Engineering Against Fracture*: Springer; 2009. p. 73-80.

[3] Selvakumar V, Palanikumar K, Palanivelu K. Studies on mechanical characterization of polypropylene/Na -MMT nanocomposites. Journal of Minerals and Materials Characterization and Engineering 2010;**9**(08):671.

[4] Luyt A, Dramićanin M, Antić Ž, Djoković V. Morphology, mechanical and thermal properties of composites of polypropylene and nanostructured wollastonite filler. *Polym Test* 2009;**28**(3):348-356.

[5] Akhtar MJ, Ahamed M, Khan M, Alrokayan SA, Ahmad I, Kumar S. Cytotoxicity and apoptosis induction by nanoscale talc particles from two different geographical regions in human lung epithelial cells. *Environ Toxicol* 2014;**29**(4):394-406.

[6] Lordan S, Kennedy JE, Higginbotham CL. Cytotoxic effects induced by unmodified and organically modified nanoclays in the human hepatic HepG2 cell line. *Journal of applied toxicology* 2011;**31**(1):27-35.

[7] Maxim LD, McConnell E. A review of the toxicology and epidemiology of wollastonite. *Inhal Toxicol* 2005;**17**(9):451-466.

[8] Njuguna J, Pielichowski K, Zhu H. Health and Environmental Safety of Nanomaterials: Polymer Nancomposites and Other Materials Containing Nanoparticles. : Elsevier; 2014.

[9] Gendre L, Marchante V, Abhyankar HA, Blackburn K, Temple C, Brighton JL. Development of CNC prototype for the characterization of the nanoparticle release during physical manipulation of nanocomposites. *Journal of Environmental Science and Health*, Part A 2016;**51**(6):495-501.

[10] Sachse S, Silva F, Zhu H, Irfan A, Leszczyńska A, Pielichowski K, et al. The effect of nanoclay on dust generation during drilling of PA6 nanocomposites. *Journal of Nanomaterials* 2012;**2012**:26.

[11] Bello D, Wardle BL, Zhang J, Yamamoto N, Santeufemio C, Hallock M, et al. Characterization of exposures to nanoscale particles and fibers during solid core drilling of hybrid carbon nanotube advanced composites. *International journal of occupational and environmental health* 2013.

[12] Starost K, Njuguna J. A review on the effect of mechanical drilling on polymer nanocomposites. In *IOP Conference Series: Materials Science and Engineering* 2014 (Vol. **64**, No. 1, p. 012031). IOP Publishing.

[13] Starost K, Frijns E, Laer JV, Faisal N, Egizabal A, Elizextea C, Blazquez M, Njuguna J. The Effect of Drilling on Nanoparticles Release from Silica and Alumina Reinforced Polyester Nanocomposites. 7th International Symposium on Nanotechnology, Occupational and Environmental Health; 2015.

[14] Irfan A, Sachse S, Njuguna J, Pielichowski K, Silva F, Zhu H. Assessment of nanoparticle release from polyamide 6-and polypropylene-silicon composites and cytotoxicity in human lung A549 cells. *Journal of Inorganic and Organometallic Polymers and Materials* 2013;**23**(4):861-870.

[15] ISO B. 527-2: 1996. Plastics—determination of tensile properties—part 2: test conditions for moulding and extrusion plastics. *British Standards Institution*. 1996 Jul:1-4.

IOP Conf. Series: Materials Science and Engineering **195** (2017) 012012 doi:10.1088/1757-899X/195/1/012012

Computational study of nanostructured composite materials for photonic crystal fibre sensors

Jincy Johny, Radhakrishna Prabhu, Wai Keung Fung
School of Engineering, Robert Gordon University
Aberdeen AB10 7GJ, United Kingdom

E-mail: r.prabhu@rgu.ac.uk

Abstract. Photonic Crystal Fibres (PCFs) developed using nanostructured composite materials provides special optical properties which can revolutionise current optical sensing technologies. The modal and propagation characteristics of the PCF can be tailored by altering their geometrical parameters and material infiltrations. A drawback of commercially available PCF is their limited operating wavelengths, which is mostly in the infrared (IR) spectral band. Nanostructured composite materials manipulates the optical properties of the PCF, facilitating their operation in the higher sensitivity near infrared (NIR) wavelength regime. Hence, there arises a need to closely investigate the effect of nanostructure and composite materials on various optical parameters of the PCF sensor. This paper presents a hexagonal PCF designed using COMSOL MULTIPHYSICS 5.1 software, with a nanostructured core and microstructured cladding. Propagation characteristics like confinement loss and mode field diameter (MFD) are investigated and compared with various geometrical parameters like core diameter, cladding hole diameter, pitch, etc. Theoretical study revealed that a nanostructured PCF experiences reduced confinement losses and also improved mode field diameter. Furthermore, studies are also carried out by infiltrating the cladding holes with composite materials (liquid crystal and glass). These simulations helped in analysing the effect of different liquid crystal materials on PCF bandwidth and spectral positions.

1. Introduction

The demand for optical fibres and fibre optic technologies increased considerably after the telecommunication revolution which ignited in the 1980's. The exceptional characteristics of optical fibres such as structural versatility and enhanced sensitivity over current techniques make them a suitable candidate for sensing applications. They have many advantages such as immunity from electromagnetic interference (EMI), electrical isolation, small size, remote sensing ability, multiplexing capability, freedom from corrosion and ability to operate in extreme environmental conditions [1]. Despite the fact that conventional optical fibres offer remarkable performance in the communication industry, it doesn't satisfy all the needs of sensing industry. The specific requirements of sensing industry include long distance real-time remote sensing with multi-point and multi-parameter sensing capability. Hence, studies are emerging on speciality fibres like fibre Bragg grating (FBG), photonic crystal fibres (PCF), doped fibres etc.

PCFs [2] are gaining popularity in recent years, owing to their specialized geometrical structure (core-air hole cladding) and unique properties, which include their guiding mechanisms and modal characteristics [3]. They produce lower optical transmission losses compared to standard optical fibres [4]. PCFs are more flexible than normal optical fibres, because it is possible to manage their properties, leading to a freedom of design [5]. Propagation characteristics of PCFs can be tuned by altering different structural or physical parameters like core diameter (ρ), cladding hole diameter (d) and pitch (Λ) in combination with the choice of material refractive index and type of crystal lattice [5-6]. All these inherent capabilities of PCFs are being exploited for different sensing applications. However, new requirements of fibre optic sensors create a need for different materials and fibre structure which can provide special optical properties. Structuring greatly boosts the performance of the fibre sensors in specific applications. Nanostructured optical materials will provide special properties which can improve the capabilities of existing optical sensing technologies [7-8]. Nanostructuring of PCF air holes is expected to reduce the fibre confinement losses and also decrease its effective mode area, which in turn improves the fibre propagation distance and range of the fibre sensor. Furthermore, addition of composite materials (liquid crystal and glass) on to the cladding holes modifies PCF transmission and polarization properties, enabling them to operate within the photonic bands having highest sensitivities. The near infrared (NIR) wavelengths of the optical spectrum are of particular interest for fibre-optic sensing applications due to their improved sensitivity and accuracy compared to other spectral regions. Another advantage of material infusion is that the properties of the PCF can be modified even after the fabrication of the PCF. The PCF air holes filled with special materials like liquid crystals changes their optical properties in response to electric or magnetic fields or light intensity. Liquid crystal materials infused into the cladding holes creates a PBG (Photonic Bandgap) effect, restricting the modes within the core region rather than leaking, facilitating a stronger sensing signal.

2. Theory

PCFs normally have two modes of operation based on their light guiding technique– index guiding and bandgap guiding [9]. Index guiding PCFs operates similar to conventional optical fibres wherein light is confined within the high index core by modified total internal reflection (M-TIR) principle [10]. On the other hand, bandgap PCFs guides light in the low index core region by reflection from the photonic crystal cladding [11].

2.1. Confinement loss

Confinement or leakage loss [12] expressed in dB/km is given by:

$$L = \frac{\left(\frac{20}{ln\,10}(2\pi)\right) Im(n_{eff})}{\lambda} \tag{1}$$

Where $Im(n_{eff})$ is the imaginary part of the effective refractive index, n_{eff} and λ is the propagating wavelength.

2.2. Effective mode area

Effective mode area [13] of the PCF is given by the equation:

$$A_{eff} = \frac{\left[\iint_{-\infty}^{\infty} |E(x,y)|^2\, dxdy\right]^2}{\iint_{-\infty}^{\infty} |E(x,y)|^4\, dxdy} \tag{2}$$

Where $E(x,y)$ is the optical mode field distribution.
Also, effective area is related to mode field diameter (MFD) by the equation:

$$A_{eff} = k_n \left(\frac{\pi}{4}\right) MFD^2 = k_n\, \pi\, w^2 \tag{3}$$

Where k_n is the correction factor and spot size, $w = MFD/2$ [14].
Mode Field diameter is approximated as,

$$MFD \approx \frac{2}{\sqrt{\pi}} \sqrt{A_{eff}} \tag{4}$$

IOP Conf. Series: Materials Science and Engineering **195** (2017) 012012 doi:10.1088/1757-899X/195/1/012012

3. Nanostructured PCF Design

Current studies are carried out by changing the size (nano size holes in the core and micro size holes in the cladding), shape (circular, elliptical) and distribution of PCF air holes. In the previous study carried out, the designed solid core PCF exhibited limited operating wavelengths [15]. Hence, inorder to achieve spectral shifting and bandwidth enhancement, core nanostructuring and also the effect of different liquid crystal materials infiltrations within the cladding holes are being investigated. PCF parameters such as confinement losses, effective area and MFD are studied as a function of normalized frequency (Λ/λ). MFD and effective area is associated with the electric field distribution within the fibre. Analysis of these parameters gives a better insight of the propagation characteristics of the PCF.

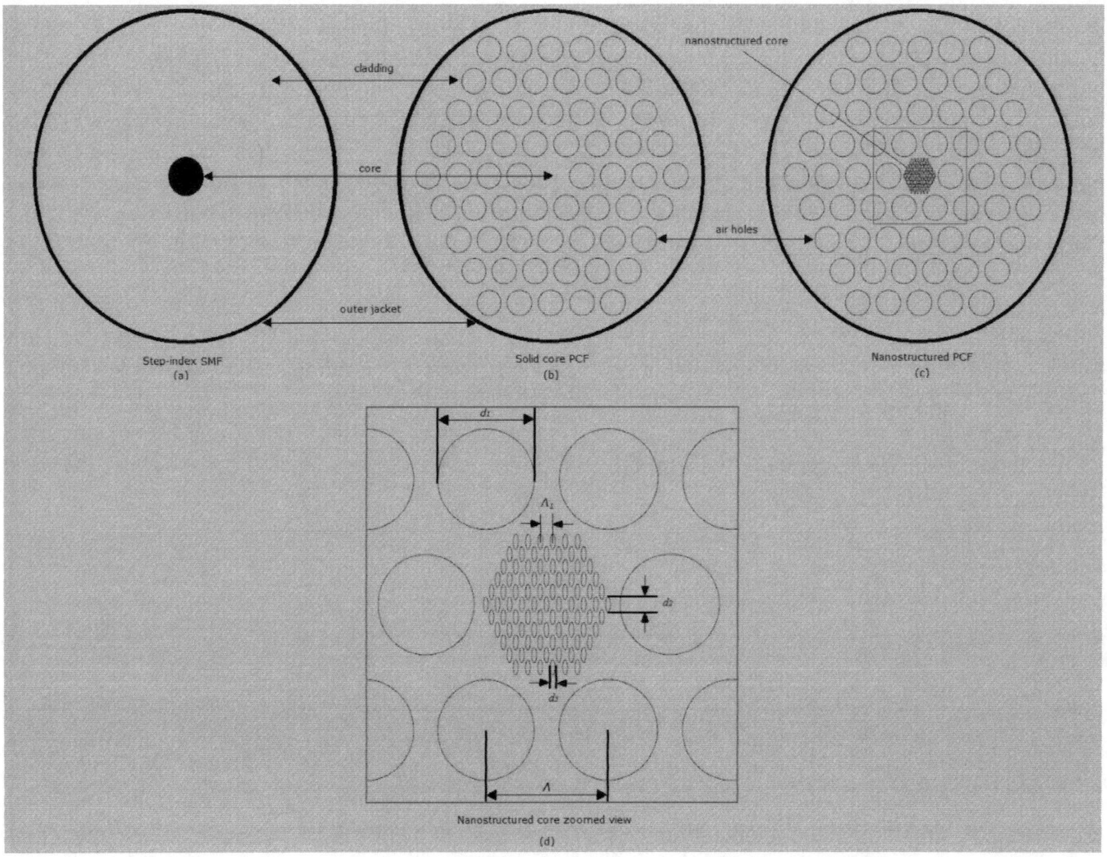

Figure 1. *Cross section of: (a) step-index SMF, (b) solid core PCF, (c) designed nanostructured PCF and (d) nanostructured core zoomed view*

Figure 1 (a) and (b) shows the cross sectional view of the structural difference between a standard step-index single mode fibre (SMF) and a solid core PCF with microstructured cladding. During the fibre modelling, the only physical parameter that needs to be taken into account in the case of SMF is the core diameter. On the other hand, while modelling a PCF, three geometrical parameters are to be considered, core diameter, cladding hole diameter and also the pitch (hole to hole distance). Figure 1 (c) and (d) shows the cross section of the designed nanostructured PCF with a four ring hexagonal lattice of circular air holes in the cladding and a five ring array of elliptical air holes in the core. Simulations were conducted on a hexagonal PCF designed using the wave optics module of COMSOL MULTIPHYSICS 5.1. The geometrical parameters of the nanostructured PCF comprises of: pitch (Λ)

which is the cladding hole center to center distance; d_1 is the diameter of the cladding air hole; d_2 and d_3 are the length of major axis and minor axis of the elliptical air holes within the core and Λ' is the hole to hole spacing between the elliptical air holes of core.

4. Simulation Results and Discussion

(a) (b)

Figure 2. *(a) Electric field mode profile of the nanostructured PCF, (b) Zoomed view of electric field mode profile in PCF core*

Figure 2 (a) and (b) shows the electric field profile of the fundamental mode obtained for the designed nanostructured PCF. From table 1 we can see that the mode confinement wavelengths can be brought down from the infrared (IR) to the near infrared (NIR) wavelengths by tuning the physical parameters of the core and cladding air holes. NIR lower cut off wavelength of 925 nm was achieved for the nanostructured PCF with design parameters: $\Lambda = 1\mu m$, $d_1 = 0.8\mu m$, $\Lambda' = 100nm$, $d_2 = 60nm$, $d_3 = 20nm$.

Table 1. Confinement wavelengths with varying physical parameters

Design	Core dimensions	Cladding dimensions	Confinement wavelengths
1	$\Lambda' = 0.1\Lambda = 0.22\mu m = 220nm$ $d_2 = 0.6\Lambda' = 0.132\mu m = 132nm$ $d_3 = 0.2\Lambda' = 0.044\mu m = 44nm$	$\Lambda = 2.2\mu m$ $d_1 = 0.8\Lambda = 1.76\mu m$	≥ 1950 nm
2	$\Lambda' = 0.1\Lambda = 0.2\mu m = 200nm$ $d_2 = 0.6\Lambda' = 0.12\mu m = 120nm$ $d_3 = 0.2\Lambda' = 0.04\mu m = 40nm$	$\Lambda = 2\mu m$ $d_1 = 0.8\Lambda = 1.6\mu m$	≥ 1775 nm
3	$\Lambda' = 0.1\Lambda = 0.15\mu m = 150nm$ $d_2 = 0.6\Lambda' = 0.09\mu m = 90nm$ $d_3 = 0.2\Lambda' = 0.03\mu m = 30nm$	$\Lambda = 1.5\ \mu m$ $d_1 = 0.8\Lambda = 1.2\mu m$	≥ 1450 nm
4	$\Lambda' = 0.1\Lambda = 0.1\mu m = 100nm$ $d_2 = 0.6\Lambda' = 0.06\mu m = 60nm$ $d_3 = 0.2\Lambda' = 0.02\mu m = 20nm$	$\Lambda = 1\mu m$ $d_1 = 0.8\Lambda = 0.8\mu m$	≥ 925 nm

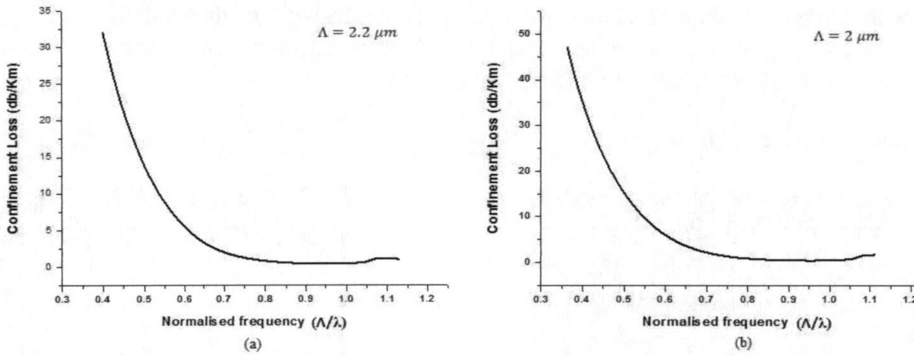

Figure 3. *Confinement loss vs normalised frequency for: (a)* $\Lambda = 2.2\mu m$; *(b)* $\Lambda = 2\mu m$

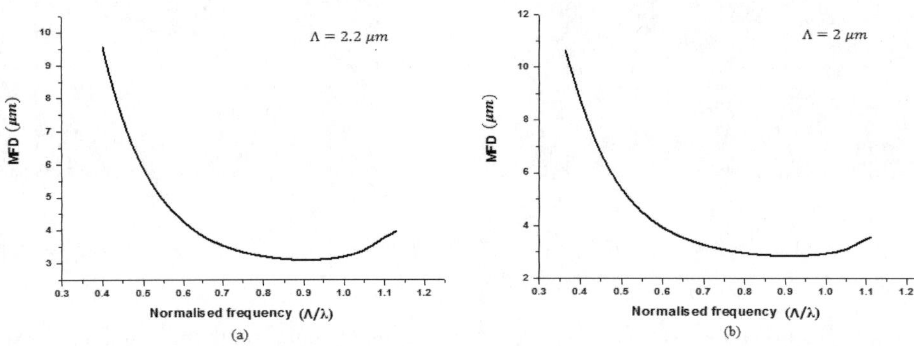

Figure 4. *MFD vs normalised frequency for: (a)* $\Lambda = 2.2\mu m$; *(b)* $\Lambda = 2\mu m$

Figure 3 depicts the variation in confinement loss (in dB/km) with respect to normalized frequency ($v = \Lambda/\lambda$) for different pitch or cladding hole spacing. It can be observed that confinement losses decreases with normalized frequency or in other words confinement losses increases with wavelength. Another advantage of bringing down the mode confinement wavelengths from the IR to NIR wavelengths is a reduction in confinement losses. Figure 4 shows the changes in MFD (in μm) with respect to normalized frequency for different pitch values. From the graph it is clear that MFD decreases with normalized frequency, or MFD experiences an increase for longer wavelengths and a decrease for shorter wavelengths. This is because, at lower wavelengths, the power density of the electric field propagating through the fibre would be higher, intensity would be lower and hence the effective mode area and MFD would be smaller. From table 2 we can see that by infiltrating different liquid crystal materials on to the cladding holes, a shifting and broadening of the spectral bands can be observed. Mode confinements in the NIR wavelengths were achieved by infiltrating the holes with liquid crystal material Cat No. 1550. To sum up, PCF nanostructuring and liquid crystal material infiltrations, reduces its losses and enhances the PCF sensor transmission distance and coverage.

Table 2. Confinement wavelengths of PCF filled with different liquid crystal materials

Sl No.	Liquid crystal material	Refractive index	Confinement wavelengths
1	K21	1.732	1650 -2000nm
2	PCH-5	1.6049	1200-1650 nm
3	Cat No. 1550	1.522	825-1100 nm

IOP Conf. Series: Materials Science and Engineering **195** (2017) 012012 doi:10.1088/1757-899X/195/1/012012

5. Conclusion

The computational study carried out by nanostructuring the PCF core and by changing the size, shape and distribution of the PCF holes resulted in a shift in its spectral band, accompanied with a reduction in confinement losses and MFD. It was found that the PCF confinement wavelengths and losses are a strong function of its structural parameters. Further simulations carried out by varying the liquid crystal material infiltrations within the PCF holes resulted in a wavelength shift from IR to NIR. Moreover, altering refractive index of the liquid crystal material filled into the cladding holes of PCF caused a shifting and broadening of the photonic bands. Through these simulations it was identified that the spectral positions and bandgaps can be tuned by nanostructuring the PCF holes and changing its material infiltrations. Hence, nanostructuring and composite material infiltrations makes it possible to have different designer wavelengths for the PCF sensors. In addition, low confinement losses and MFD improves the signal power of the sensor, which in turn enhances its range and propagation distance.

6. References

[1] Lee B. Review of the present status of optical fiber sensors. *Optical Fiber Technology.* 2003; 9(2):57-79.

[2] Knight J, Birks T, Russell PSJ, Atkin D. All-silica single-mode optical fiber with photonic crystal cladding. *Optics Letters.* 1996; 21(19):1547-1549.

[3] Buczynski R. Photonic crystal fibers. *Acta Physica Polonica Series A.* 2004; 106(2):141-168.

[4] Knight JC. Photonic crystal fibres. *Nature.* 2003; 424(6950):847-851.

[5] Pinto AM, Lopez-Amo M. Photonic crystal fibers for sensing applications. *Journal of Sensors.* 2012; 2012.

[6] Ahmed K, Morshed M. Design and numerical analysis of microstructured-core octagonal photonic crystal fiber for sensing applications. *Sensing and Bio-Sensing Research.* 2016; 7:1-6.

[7] Zhang Y. *Physical properties investigation of nanostructured materials and their applications.* : ProQuest; 2008.

[8] Paquet C, Kumacheva E. Nanostructured polymers for photonics. *Materials Today.* 2008; 11(4):48-56.

[9] Pal BP. *Guided wave optical components and devices: basics, technology, and applications.* : Academic press; 2010.

[10] Wehrspohn RB, Kitzerow H, Busch K. *Nanophotonic Materials: Photonic Crystals, Plasmonics, and Metamaterials.* : John Wiley & Sons; 2008.

[11] Xiao L, Jin W, Demokan M. Photonic crystal fibers confining light by both index-guiding and bandgap-guiding: hybrid PCFs. *Optics express.* 2007; 15(24):15637-15647.

[12] Sakai J, Niiro H. Confinement loss evaluation based on a multilayer division method in Bragg fibers. *Optics express.* 2008; 16(3):1885-1902.

[13] Miyagi K, Namihira Y, Razzak SA, Kaijage SF, Begum F. Measurements of mode field diameter and effective area of photonic crystal fibers by far-field scanning technique. *Optical review.* 2010; 17(4):388-392.

[14] Mortensen NA. Effective area of photonic crystal fibers. *Optics Express.* 2002; 10(7):341-348.

[15] Johny J, Prabhu R, Fung WK. Investigation of structural parameter dependence of confinement losses in PCF–FBG sensor for oil and gas sensing applications. *Optical and Quantum Electronics.* 2016; 48(4):1-9.

IOP Conf. Series: Materials Science and Engineering **195** (2017) 012013 doi:10.1088/1757-899X/195/1/012013

Investigations on mechanism of self-healing and cavity filling in case of steel inoculated with seashell powder

Parvaiz Habibullah[1], Ghazala Siddiqui[2], Yasir Saleem[3]
[1]Interdisciplinary Res. in Met. & Mat. for Adv. Appl., Lahore, Pakistan
[1]Dr. Parvaiz Habibullah, 22-Davis Road, Lahore – Pakistan
[2]Director, Centre of Excellence in Marine Biology, University of Karachi, Karachi, Pakistan
[3]Senior Scientific Officer, Food & Biotechnology Research Centre, PCSIR, Labs. Lahore, Pakistan

E-mail: drpervaiz.h@gmail.com; www.drpervaiz-h.org

Abstract. Seashells are comprised largely of a brittle ceramic material (calcite, the stuff of chalk) in the form of microscopic slates. Shells, such as the abalones, reinforced with a kind of protein mortar in efforts to evaluate on how seashells repair themselves. We have identified the protein and mechanism on how the protein mortar stretches itself into ligaments that bridge the gap, with the help of scanning electron microscope (SEM).

Introduction

Recently scientists have come up with some stuffs that can repair themselves. One of the models for this research comes from steel inoculated with seashells powder. Seashells are comprised largely of a brittle ceramic material (calcite, the stuff of chalk) in the form of microscopic slates. Shells, such as the abalones, are reinforced with a kind of protein mortar. Scanning electron microscope pictures of calcite prisms of *Mytilus edulis* shell and of lime crystals of surface turbo snail, at 1500×, are shown respectively in (a) and (b).

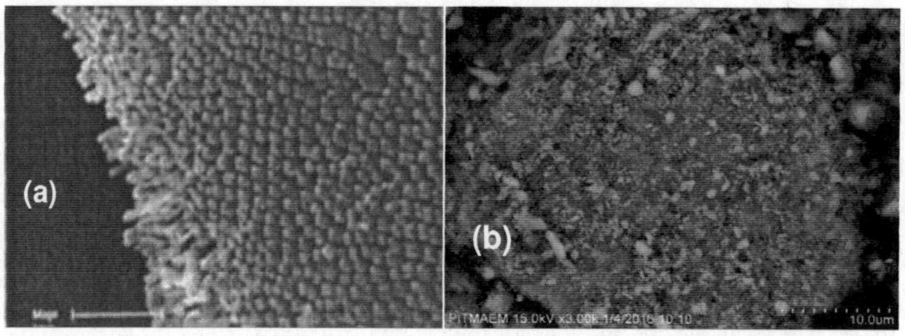

Experiments

We utilized seashell powder, composition: moisture = 0.280%, L.O.I @950 °C (as CO_2) = 39.92%, SiO_2 = 5.980%, Al_2O_3 = 2.558%, Fe_2O_3 = 0.518%, CaO = 47.86%, MgO = 2.232%, Na_2O = 0.582%, K_2O = 0.0356%, $PO_4)^{-3}$ = 0.030%, Estimated crude protein = 1.12%, [SEM image of powder at 3000× is shown in (c)]. Low carbon steel scrap was melted at 1580 – 1600°C in induction furnace and was tapped in ladle containing dry oyster sea shell powder, at its bottom. 350gms powder was added in 30-35kg steel. Test bars of 12″ length and 1¼″ diameter were cast.

1) The samples of the test bar made were observed in the optical and scanning electron microscope (SEM). One of the image at 15000×, shown in (fig. 2), is confirmed as pearlitic steel structure. EDX analysis results of this structure shows 28.04% C and 71.40% Fe,

2) Scanning Electron Microscope (SEM) images and EDX analysis of a pore at the grain boundary of the samples reveals that it contains 100% Fe (fig. 3)

Results and Conclusions

The present investigations on mechanism of self-healing and cavity filling, in case of steel inoculated with seashell powder we concluded as follows:

i. The numerous research work has confirmed that seashell are comprised largely of a brittle ceramic material (calcite, the stuff of chalk) in the form of microscopic slates and they are reinforced with a kind of protein mortar. The oyster seashell powder, used for inoculation, contains total $CaCO_3$ almost 88% and 1.12% crude protein of characteristic bands of 70Kda and 40Kda (fig.1).

ii. In seashells, protein layer immediately intercepts cracks that begin in any calcium carbonate layer. The crack thus requires still more energy before it can propagate through succeeding calcium carbonate layers, thus improving their resistance to fracture.

iii. When cracks begin to form in a metallurgical material, this protein mortar stretches itself into ligaments that bridge the gap and pull the two sides back together. About protein, it is a confirmed fact that if they degenerate, they can regenerate and reunite after some time, provided they were not burnt to ashes. In case of steel, under the effect of heat, the protein changes shape to clamp the sides of the crack shut.

iv. Scanning Electron Microscope (SEM) images and EDX analysis of a pore at the grain boundary of the samples reveals that it contains 100% Fe. It indicates the adjoining material fills that anypore appearing.

v. None of the SEM image and EDX analysis of the samples show any presence of Ca, Sr and Mg. Ca imparts greater fluidity and has high vapour pressure since its boiling point is lower than temperature of liquid steel. Ca and other elements such as Sr, Mg present in minute quantity in seashells are consumed during inoculation of steel with sea shell powder.

Sea shell protein sample showing two major bands and some minor bands

Observations: Samples were labeled as **M:** Molecular weigh marker; **spill**: standard **BSA** spills; **BSA**: Bovine serum albumin (used as standard protein of known molecular weight); **shell**:

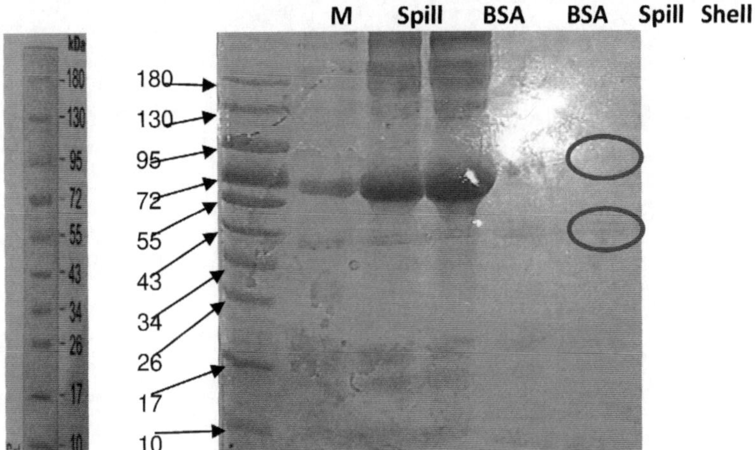

Fig. 1 SDS PAGE (Sodium Dodecyl Sulfate Polyacrylamide Gel Electrophoresis) analysis was carried out on Sea shell proteins on 12 % gel. Sea shell samples. Sea shell protein sample shows

IOP Conf. Series: Materials Science and Engineering **195** (2017) 012013 doi:10.1088/1757-899X/195/1/012013

two major bands and some minor bands. Characteristic bands of 70 Kda and 40 Kda were observed in it.

Schematic of SDS analysis of seashell protein compared to BSA*

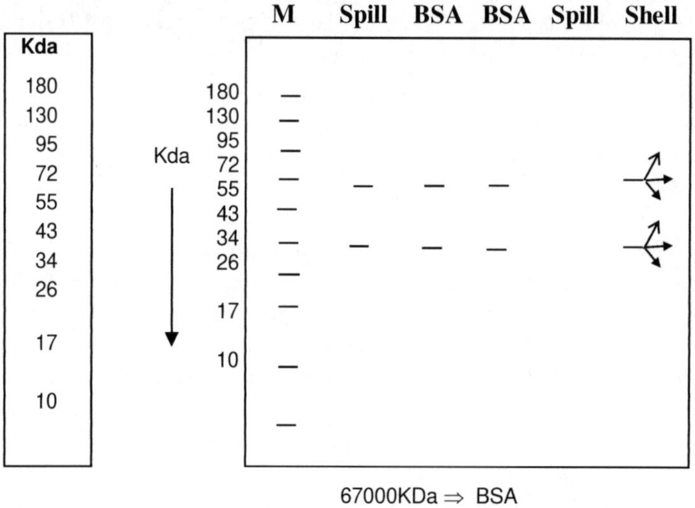

Sea shell protein sample showing two major bands and some minor bands

Observations: Samples were labelled as **M:** Molecular weigh marker; **spill**: standard **BSA** spills; **BSA**: Bovine serum albumin (used as standard protein of known molecular weight); **shell**:

IOP Conf. Series: Materials Science and Engineering **195** (2017) 012013 doi:10.1088/1757-899X/195/1/012013

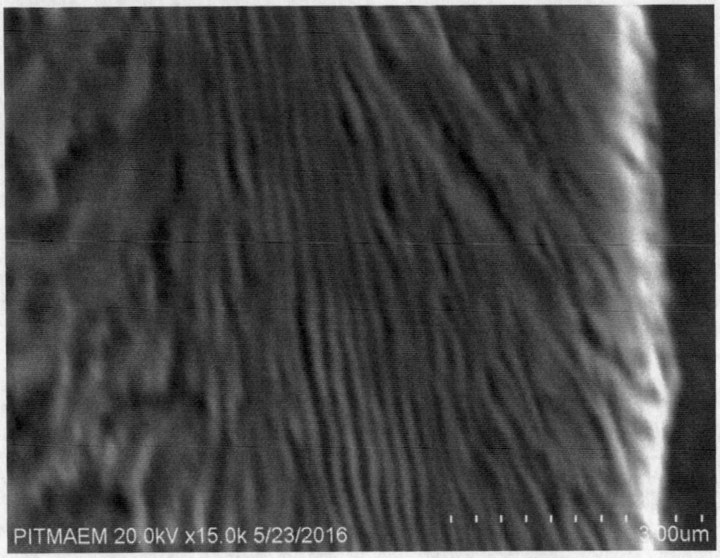

Pearlitic Structure 15000×

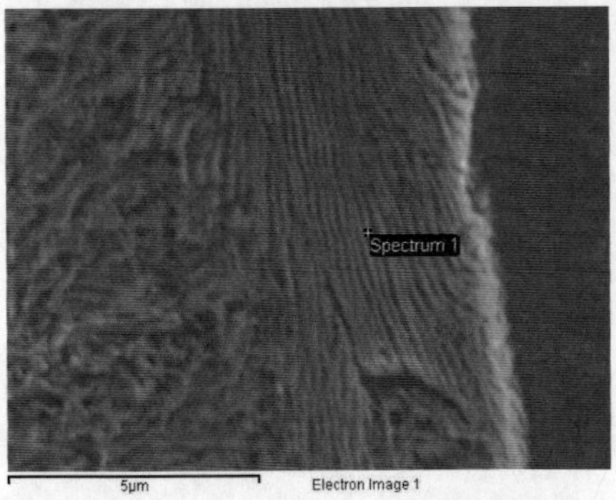

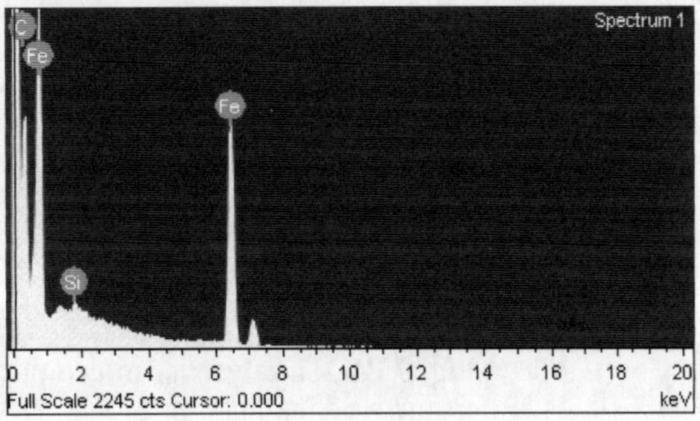

Element	Weight%	Atomic%
C	28.04	64.27
Si	0.56	0.54
Fe	71.40	35.19
Totals	100.00	

Fig. 2. Pearlitic structure and its EDX analysis

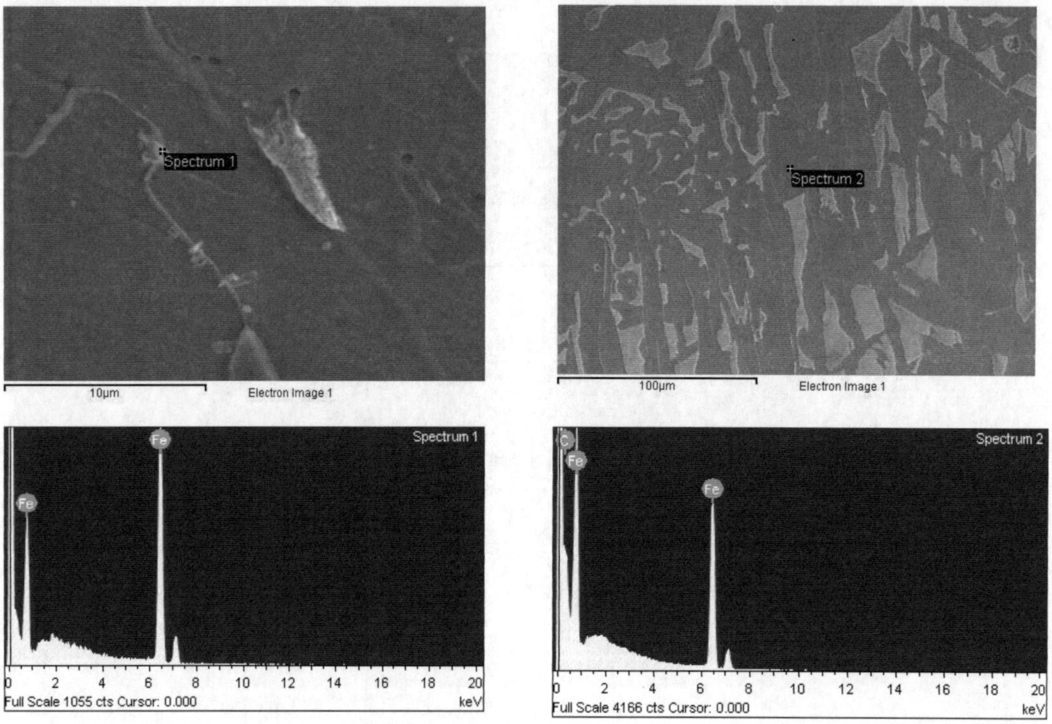

Element	Weight%	Atomic%
Fe	100.00	100.00
Totals	100.00	

Fig. 3 EDX analysis of micro pore at grain boundary

References

1. V. Von, K. Stemmer, Shell formation and microstructure of the ocean quahog Arctica islandica: Does ocean acidification matter? Dissertation zur erlangung des Grades eines doctors der Naturwissenschaften, University of Bremen, 2013.
2. K. Stemmer, T. Brey, M. Beutler, B. Schalkhausser, D. de Beer. In situe measurements of pH, Ca^{2+} and DIC dynamics within the extrapallial fluid of the ocean quahog Arctica islandica, Journal of Experimental Marine biology and Ecology, 2013.
3. B. Schalkhausser, C. Bock, K. Stemmer, T. Brey, H. O Portner, G. Lannig. Impact of ocean acidification on escape performance of the king scalolop, Pecten maximus, from Normay, Mar Biol. DOI 10.1007/s00227-012-2057-8, 2012.
4. K. Shiral, N. Takahata, H. Yamamoto, T. Omata, T. Sasaki and Y. Sano, Novel analytical approach to bivalve shell biogeochemistry: A case study of hydrothermal mussel shell. Geochemical Journal, Vol. 42, pp. 413 to 420, 2008

5. S.R. White, N.R. Sottos, P.H. Geubelle, J.S. Moore, M.R. Kessler, S.R. Sriram, E.N. Brown and S. Viswanathan, "Autonomic Healing of Polymer Composites", Nature, 2001, 409(6822), 794-797.
6. J.A. Pechenik, Biology of the Inverterbrates, 4th Ed. McGraw Hill, 2000
7. Ivan Amato, Self-healing steel? Intelligent bridges? Research into new materials could well result in the stuff that dreams are made of.... Time, Jan. 1998, p.69-72.
8. C.M. Dry, "Smart Building Materials Which Prevent Damage and Repair Themselves, in Smart Materials Fabrication and Materials for Micro-Electro-Mechanical Systems, Symposium held April 28-30, 1992, San Francisco, CA, Vol. 276 in series Materials research Society Proceedings, MRS Philadelphia, PA 331, (1992).
9. R. P. Wool, Self-healing materials: a review. The Royal Society of Chemistry, 2008, 4, 400-418.
10. Turkdogan E T, Steel research, 62, 1991, 379.
11. R. T. Abbott, Kingdom of the seashell, Bonanza Books, New York, 1982.
12. *Chick H & Martin C.J, The heat coagulation of proteins. Part-VI. The conditions controlling the agglutination of proteins already acted upon by hot water. Lister Institute of Preventive Medicine.
13. Wikipedia "seashell"
14. *Calcium in steels. pdfcrowd.com
15. Coastalcare.org.educate/seashells